Advances in the Preparation, Properties and Application of Polyurethane, Cellulose and Their Composites

Advances in the Preparation, Properties and Application of Polyurethane, Cellulose and Their Composites

Editors

Hui Zhao
Yang Liu
Yan Jiang

Basel • Beijing • Wuhan • Barcelona • Belgrade • Novi Sad • Cluj • Manchester

Editors

Hui Zhao
Guangxi University
Naning
China

Yang Liu
Guangxi University
Naning
China

Yan Jiang
Guangxi University
Naning
China

Editorial Office
MDPI
St. Alban-Anlage 66
4052 Basel, Switzerland

This is a reprint of articles from the Special Issue published online in the open access journal *Polymers* (ISSN 2073-4360) (available at: http://www.mdpi.com).

For citation purposes, cite each article independently as indicated on the article page online and as indicated below:

Lastname, A.A.; Lastname, B.B. Article Title. *Journal Name* **Year**, *Volume Number*, Page Range.

ISBN 978-3-7258-0183-1 (Hbk)
ISBN 978-3-7258-0184-8 (PDF)
doi.org/10.3390/books978-3-7258-0184-8

© 2024 by the authors. Articles in this book are Open Access and distributed under the Creative Commons Attribution (CC BY) license. The book as a whole is distributed by MDPI under the terms and conditions of the Creative Commons Attribution-NonCommercial-NoDerivs (CC BY-NC-ND) license.

Contents

Yuhan Zhu, Fei Guo, Jing Li, Zhen Wang, Zihui Liang and Changhai Yi
Development of a Novel Energy Saving and Environmentally Friendly Starch via a Graft Copolymerization Strategy for Efficient Warp Sizing and Easy Removal
Reprinted from: *Polymers* **2024**, *16*, 182, doi:10.3390/polym16020182 . . . . **1**

Fei Guo, Hui Ma, Bin-Bin Yang, Zhen Wang, Xiang-Gao Meng, Jian-Hua Bu and Chun Zhang
Rigidity with Flexibility: Porous Triptycene Networks for Enhancing Methane Storage
Reprinted from: *Polymers* **2024**, *16*, 156, doi:10.3390/polym16010156 . . . . **14**

Yajing Wang, Juan Li, Ru Lin, Dianrun Gu, Yuanfang Zhou, Han Li and Xiangdong Yang
Recommended Values for the Hydrophobicity and Mechanical Properties of Coating Materials Usable for Preparing Controlled-Release Fertilizers
Reprinted from: *Polymers* **2023**, *15*, 4687, doi:10.3390/polym15244687 . . . . **25**

Yongxing Zhou, Wenbo Yin, Yuliang Guo, Chenni Qin, Yizheng Qin and Yang Liu
Green Preparation of Lightweight, High-Strength Cellulose-Based Foam and Evaluation of Its Adsorption Properties
Reprinted from: *Polymers* **2023**, *15*, 1879, doi:10.3390/polym15081879 . . . . **42**

Yiqi Chen, Yujie Duan, Han Zhao, Kelan Liu, Yiqing Liu, Min Wu and Peng Lu
Preparation of Bio-Based Foams with a Uniform Pore Structure by Nanocellulose/Nisin/Waterborne-Polyurethane-Stabilized Pickering Emulsion
Reprinted from: *Polymers* **2022**, *14*, 5159, doi:10.3390/polym14235159 . . . . **57**

Zhenqi Zhou, Chunlin Jiao, Yinna Liang, Ang Du, Jiaming Zhang, Jianhua Xiong, et al.
Study on Degradation of 1,2,4-TrCB by Sugarcane Cellulose-TiO_2 Carrier in an Intimate Coupling of Photocatalysis and Biodegradation System
Reprinted from: *Polymers* **2022**, *14*, 4774, doi:10.3390/polym14214774 . . . . **69**

Xueyan Liu, Chuxing Zhu, Kang Yu, Wei Li, Yingchun Luo, Yi Dai and Hao Wang
Accurate Determination of Moisture Content in Flavor Microcapsules Using Headspace Gas Chromatography
Reprinted from: *Polymers* **2022**, *14*, 3002, doi:10.3390/polym14153002 . . . . **85**

Zhou Wan, Chunlin Jiao, Qilin Feng, Jue Wang, Jianhua Xiong, Guoning Chen, et al.
A Cellulose-Type Carrier for Intimate Coupling Photocatalysis and Biodegradation
Reprinted from: *Polymers* **2022**, *14*, 2998, doi:10.3390/polym14152998 . . . . **95**

Yaohui Wang, Long Li, Gege Cheng, Lanfu Li, Xiuyu Liu and Qin Huang
Preparation and Recognition Properties of Molecularly Imprinted Nanofiber Membrane of Chrysin
Reprinted from: *Polymers* **2022**, *14*, 2398, doi:10.3390/polym14122398 . . . . **110**

Gege Cheng, Wenwen Li, Long Li, Fuhou Lei, Xiuyu Liu and Qin Huang
Removing Calcium Ions from Remelt Syrup with Rosin-Based Macroporous Cationic Resin
Reprinted from: *Polymers* **2022**, *14*, 2397, doi:10.3390/polym14122397 . . . . **125**

Article

Development of a Novel Energy Saving and Environmentally Friendly Starch via a Graft Copolymerization Strategy for Efficient Warp Sizing and Easy Removal

Yuhan Zhu, Fei Guo, Jing Li, Zhen Wang, Zihui Liang * and Changhai Yi *

National Local Joint Laboratory for Advanced Textile Processing and Clean Production, State Key Laboratory of New Textile Materials and Advanced Processing Technologies, Wuhan Textile University, Wuhan 430073, China
* Correspondence: zhliang@wtu.edu.cn (Z.L.); ych@wtu.edu.cn (C.Y.)

Abstract: Warp sizing is a key process in textile production. However, before the yarn/fabric finishing, such as dyeing, the paste adhering to the warp must be eliminated to ensure optimal dyeing properties and the flexibility of the fabric. Therefore, the sizing will often consume a lot of energy and produce a lot of industrial wastewater, which will cause serious harm to the environment. In this study, we have developed an energy saving and environmentally friendly starch-based slurry by modifying natural starch with acrylamide. The paste has excellent viscosity stability and fiber adhesion, and exhibits excellent performance during warp sizing. In addition, the slurry has good water solubility at 60–70 °C, so it is easy to desize at low temperatures. Because of this, the sizing of the warp can be deslimed directly from the yarn during subsequent washing processes. This work can not only reduce some costs for the textile industry, but also achieve the purpose of energy conservation and emission reduction.

Keywords: starch; copolymerization; warp sizing; desizing; energy-saving and environmental-friendly

Citation: Zhu, Y.; Guo, F.; Li, J.; Wang, Z.; Liang, Z.; Yi, C. Development of a Novel Energy Saving and Environmentally Friendly Starch via a Graft Copolymerization Strategy for Efficient Warp Sizing and Easy Removal. *Polymers* **2024**, *16*, 182. https://doi.org/10.3390/polym16020182

Academic Editor: Fengwei (David) Xie

Received: 27 September 2023
Revised: 18 November 2023
Accepted: 24 November 2023
Published: 8 January 2024

Copyright: © 2024 by the authors. Licensee MDPI, Basel, Switzerland. This article is an open access article distributed under the terms and conditions of the Creative Commons Attribution (CC BY) license (https://creativecommons.org/licenses/by/4.0/).

1. Introduction

It is well known that the warp yarn will be repeatedly subjected to a series of mechanical forces during the weaving process, so that the phenomenon of frequent breakage of the warp will result in low weaving efficiency [1,2], Therefore, warp sizing is a critical process in textile production, which forms a film on the surface of the yarn to improve the mechanical properties of the warp, protecting the yarn so it is not easy to break. As a renewable natural polymer compound in nature, starch has become an important textile sizing material due to its availability, low price and biodegradability [3,4]. However, starch molecular chains are composed of cyclic glucose residue groups, which leads to poor flexibility of macromolecules, resulting in the sizing films exhibiting brittle and hard properties [5,6]. These drawbacks of natural starch seriously limit its application in textile warp sizing. Therefore, developing a new starch-based slurry that can meet the environmental requirements of warp sizing in the textile industry is an important issue and needs to be addressed [7].

Considerable efforts to improve the sizing properties of starch-based slurry to meet the requirements of warp sizing have mainly focused on oxidation [8], hydrolysis [9], graft copolymerization [10,11] and crosslinking [12]. Among the various strategies, graft copolymerization is promising approach for improving the sizing properties of starch and expanding its range of applications [13]. Meshram et al. proposed styrene (ST), butyl acrylate (BA) and methyl methacrylate (MMA) as monomers and ferrous ammonium sulfate–hydrogen peroxide initiated reaction, the optimum monomer concentration and initiator concentration were investigated. ST/BA and ST/MMA grafted starch showed superior tensile strength for cotton sizing [14]. Zhu et al. reported that acid starch (ATS) and 2-acryloxyethyl trimethyl ammonium chloride (ATAC) initiated grafting copolymerization in the ferrous ion redox system, and found that grafted starch with the best grafting percentage of 7.5% significantly improved the adhesion properties of starch with cotton fiber and polyester fiber [15]. Zha et al. studied the grafting parameters of starch and polyacrylic acid (PAA) under different conditions, using ammonium cerium nitrate as initiator, and observed that the abrasion resistance of the sizing yarns was effectively improved, while maintaining a better desizing effect [16]. Zhu et al. employed a series of maleate starches with different degree of substitution (DS) values prepared by treating maleates with maleic anhydride, and the starch films had better elongation and breaking strengths and greater bending resistance after employing low levels of maleation and sulfosuccinic acidification [17]. Djordjevic et al. successfully found that azobisisobutyronitrile triggered a graft polymerization with starch hydrolyzed by hydrochloric acid, and the grafted starch was used for sizing of cotton yarns with a more homogeneous distribution of the slurry in the fiber [18].

Previous researchers have made important contributions to the modification of natural starch-based slurry, which has great significance in promoting the development of warp sizing [19]. Most of the research focuses on the type of initiator, the concentration of monomer, the type of monomer and the conditions of graft copolymerization reaction to improve the grafting percentage, grafting efficiency and sizing performance. However, the biodegradability and desizing properties of starch-based slurry are very important for industrial production. As mentioned above, the slurry is used to improve the warp weave. When the fabric is finished, a desizing process is also needed to remove it from the surface of the warp. The common desizing methods are acid hydrolysis and enzyme hydrolysis, depending on the composition of the slurry [20]. During the process, a large amount of energy is often consumed and a large amount of industrial wastes are generated, which causes serious harm to the environment [21,22]. In this study, we developed an energy saving and environmentally friendly starch-based slurry (abbreviated to St-AM) via modifying natural starch (St) with acrylamide (AM), which significantly enhances viscosity stability, coagulation resistance and fiber adhesion of the St-AM slurry with excellent warp sizing performance. Moreover, the St-AM slurry exhibits excellent water solubility at 60–70 °C, so the slurry has excellent sizing performance and easy desizing at low temperatures. Because of this, the warp can be receded directly after sizing during subsequent washing processes, thus reducing the traditional desizing process. Overall, it will not only reduce some costs for the textile industry, but also achieve the purpose of energy conservation and emission reduction.

2. Materials and Methods

2.1. Materials

Pure cotton roving (672 tex) was used to determine fiber adhesion. Pure cotton yarn (29 tex) was supplied by Hubei Deyongsheng Textile Company Limited (Shishou, China). Natural potato starch (industrial grade) and acrylamide were purchased from Aladdin Industrial Company (Shanghai, China). Other reagents such as ammonium persulfate (APS), methanol, ethanol, glacial acetic acid and N, N-dimethylformamide were provided by Sinopharm Chemical Reagent Co. (Shanghai, China).

2.2. Preparation of St-g-PAM

Prior to graft copolymerization, potato starch was subjected to acid digestion to provide a control sample (ACS) for subsequent work. An amount of acid-dissolved starch was weighed and configured into a 4.3% starch suspension. This suspension was transferred into a four-necked flask equipped with a thermometer, condenser tube, dropping funnel and nitrogen conduit, and pasted at 65 °C for 30 min. Subsequently, when cooled down to 50 °C, the initiator, ammonium persulfate (for 4% of the mass of the dry starch) was added slowly dropwise to pre-react with the starch macromolecules for 20 min. A certain amount of acrylamide solution (configured as 10% aqueous) was then added slowly dropwise by dropping the starch with the use of a dropping funnel, so that the dropwise addition was completed within 30–40 min and then the mixture was stirred for 3 h. Finally, 50 mL methanol was added to the mixture to terminate the polymerization reaction, then the mixture was precipitated, washed with ethanol 3–4 times, and filtered for collection. The grafted samples were marked as St-g-PAM 1 (starch to acrylamide mass ratio 2:1), St-g-PAM 2 (starch to acrylamide mass ratio 1:1), St-g-PAM 3 (starch to acrylamide mass ratio 1:1.5), and St-g-PAM 4 (starch to acrylamide mass ratio 1:2). The crude product obtained was wrapped in filter paper and extracted at reflux in a mixed solvent of glacial acetic acid (AC) and N,N-dimethylformamide (1:1, v/v) for three siphon cycles. The product was then washed three times with ethanol, and the washed product was dried in a vacuum drying oven for 12 h to remove the homopolymer in the reaction.

2.3. Characterization

Monomer conversion refers to the proportion of monomers converted to polymers as a percentage of incorporated monomers [23]. The grafting percentage refers to the mass of copolymer grafted onto the molecular chain of starch as a proportion of the mass of starch, and the grafting efficiency refers to the mass of grafted copolymer as a proportion of the total mass of the initially invested monomer and starch [24]. They are calculated according to Equations (1)–(3).

$$\text{Monomer conversion ratio}(\%) = \frac{m_1 - m_2}{m_1} \times 100 \qquad (1)$$

$$\text{Grafting percentage }(\%) = \frac{m_3 - m_4}{m_4} \times 100 \qquad (2)$$

$$\text{Grafting efficiency }(\%) = \frac{m_3}{m_1 + m_4} \times 100 \qquad (3)$$

where m_1, m_2, m_3 and m_4 are the masses of monomer used, residual monomer, graft copolymer onto the molecular chain of starch and starch, respectively. In the other words, m_1 is the mass of acrylamide weighed in the grafting reaction, m_4 is the mass of starch added in the grafting reaction, m_3 is the mass of grafted starch produced after the grafting reaction, and m_2 is the mass of acrylamide not involved in the grafting reaction, which is equal to the value of $m_1 - (m_3 - m_4)$.

The viscosity of the slurry was measured using a Wu-type viscometer [25]. Firstly, the starch aqueous suspension with a concentration of 6% was configured, and then heated at 95 °C for 1 h to prepare a starch paste. The viscosity was recorded every 30 min for five consecutive times. The viscosity stability was calculated according to Equation (4).

$$V_s\ (\%) = (1 - \frac{\eta_{max} - \eta_{min}}{\eta_{t60}}) \times 100 \qquad (4)$$

where η_{max} and η_{min} denote the maximum and minimum viscosity recorded over a period of time, respectively, and η_{t60} refers to the viscosity recorded at 60 min. The adhesion of pulp to fibers was investigated using the FZ/T 15001-2017, and the adhesion properties of pulp to fibers were determined according to the literature [26]. Tensile tests were carried out on an Instron universal material testing machine at a speed of 50 mm/min and an initial clamping distance of 100 mm, and each specimen was measured 20 times to take the average value.

FTIR spectra were obtained using a Perkin-Elmer spectrometer with a test wavelength range of 3000 to 500 cm^{-1}. XRD was tested using an Empyrean-type X-ray diffractometer with a scanning range of 10–80° and a scanning time of 3 min. A scanning electron microscope (SEM) (JSM-7001F, JEOL) was used to analyze the microstructure and chemical composition of the samples. Thermogravimetry was measured using an NETZSCH TG209F1-Nevio-Thermogravimetric Analyzer (Selb, Germany), with a temperature range of 30–650 °C and a temperature increase rate of 10 °C/min. Biochemical oxygen demand (BOD$_5$) was determined according to the Italian standard method

IRSA-CNR 29/2003-5120-B2. A volume of the eluate to be tested was placed in a Winkler bottle (volume = 300 mL). The bottle was then filled with dilution water saturated in oxygen and containing bacterial inoculum and the nutrients required for biological growth. The bottle was stored in the dark at a temperature of 20 °C for 5 d. The oxygen concentration in the bottle before and after 5 d of incubation was measured by adding to the solution manganese sulfate and potassium iodide in sodium azide and titrating the residue iodine with sodium thiosulphate. The transmittance was determined using a UV-101 spectrometer with distilled water as the reference sample (the transmittance was 100%) at wavelengths ranging from 200 nm to 800 nm. The water sample was extracted to measure chemical oxygen demand (COD_{cr}) with a HACH COD reactor, which was expressed as COD_{Cr} (potassium dichromate as oxidant).

3. Results and Discussion

3.1. Grafting Process and Mechanism

Figure 1 depicts the chemical mechanism of the graft copolymerization reaction between starch and acrylamide. The water-soluble initiator APS dissociates at a suitable temperature to produce a pair of initiating radicals (SO_4). A portion of these initiating radicals diffuse in water, taking hydrogen atoms from the hydroxyl groups of polysaccharides and from the alkyl groups of monomers to create active site initiation reactions in the starch backbone (step 1). The subsequent chain growth stage (step 2) involves two reactions in which a single unit of acrylamide monomer is added to the starch backbone (1), and then the length of the grafted chain is increased by attaching more acrylamide monomers to the starch backbone (2). The chain termination stage (step 3), in which two growing chains are coupled together, terminates the growth of the latter chain. The grafted starch samples were prepared in aqueous medium using starch and acrylamide as raw materials and the entire synthesis process is shown in Figure S1.

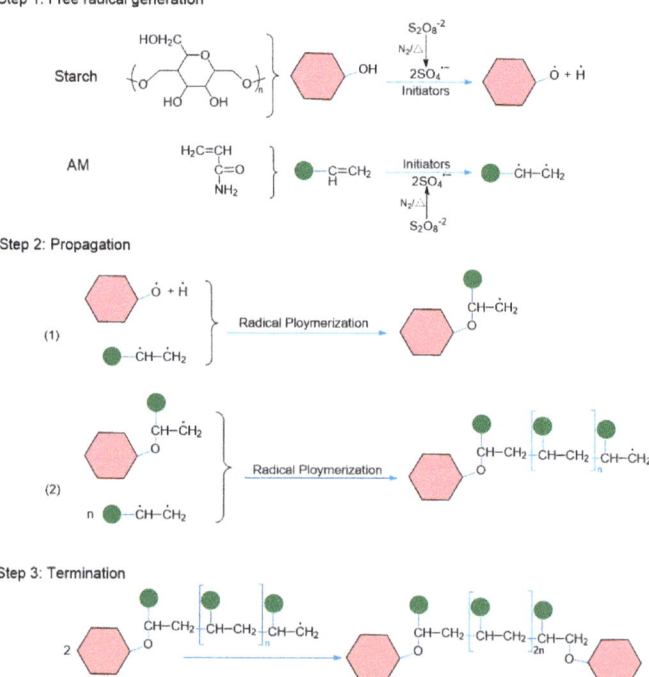

Figure 1. Synthesis process of grafted starch slurry.

3.2. Characterization of St-g-PAM

As shown in Figure 2a, we evaluated the effect of different mass ratios of St-g-PAM on the grafting percentage and grafting efficiency. The results showed that with the increase of AM monomer dosage, more monomers were attached to the free radicals on the starch backbone in the system, and thus the grafting percentage increased from 0 to 64.6%. On the other hand, the grafting efficiency of St-g-PAM showed a decreasing tendency due to the increase of the dosage of AM monomers, and as the grafting reaction continued, the active sites on the starch granules were continuously occupied, and thus there were not enough sites for the introduction of new PAM branching. In addition, the monomer-to-polymer conversions were all over 97%, indicating that most of the monomers had already undergone polymerization reactions to polymers. The detailed grafting parameters are listed in Table S1.

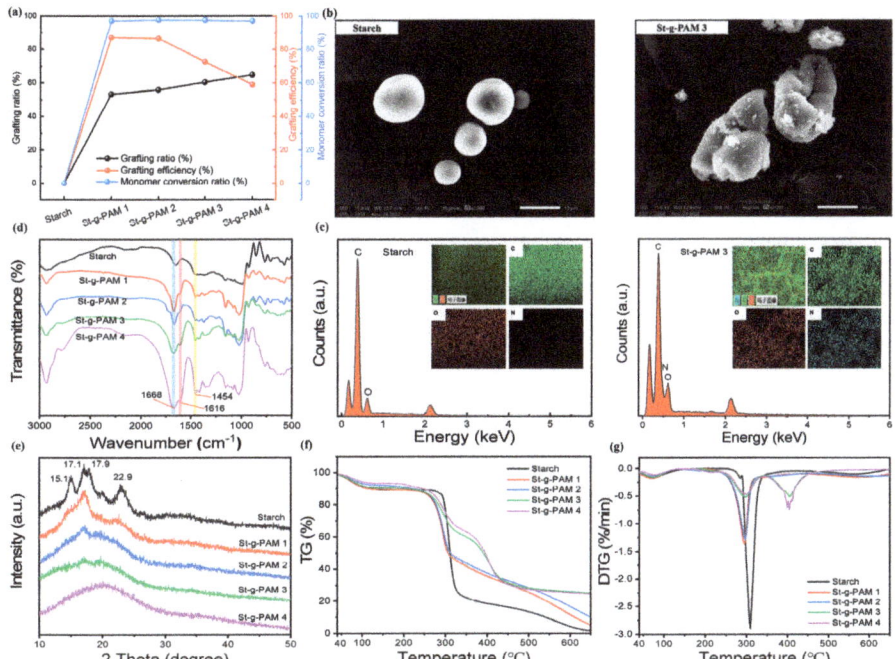

Figure 2. (**a**) Grafting parameters of starch and St-g-PAM with different mass ratios. (**b**) SEM images and (**c**) EDS spectrogram of starch and St-g-PAM 3. (**d**) FT-IR spectra and (**e**) XRD image of starch and St-g-PAM with different mass ratios. (**f**) TGA and (**g**) DTG curves of starch and St-g-PAM with different mass ratios.

The particle morphology of starch before and after modification was investigated using scanning electron microscopy (SEM), as shown in Figures 2b and S2. The surface of the granules of natural starch was smooth and had an elliptical structure, and the structure of St-g-PAM granules changed significantly after graft copolymerization. Compared with the pristine starch, the grafted starch particles were more irregular, the surface was rough, and all of them showed different degrees of depression, and these phenomena indicated that the graft modification occurred not only on the surface of the particles but also inside the particles. The EDS technique has become an important method for determining the chemical composition of polymeric materials [27]. The energy dispersive X-ray (EDX) spectroscopy confirms the main elements of the starch and St-g-PAM 3, as shown in Figure 2c. The pristine starch exhibited C and O elements, which corresponds to the composition of natural starch. In contrast, the presence of a new N element on the surface of St-g-PAM

3 indicates that the PAM side chain was successfully grafted onto the starch. In addition, the corresponding mapping of the starch and St-g-PAM 3 (the insets of Figure 2c) further confirmed the findings. To ensure the robustness of these observations, similar analyses were conducted on the St-g-PAM 1, St-g-PAM 2, and St-g-PAM 4, as shown in Figure S3. As the amount of PAM added increased, the amount of elemental N increased. The elemental composition of N indicates that the PAM side chain was successfully grafted onto the starch.

The FT-IR spectra of starch and St-g-PAM with different raw material ratios are shown in Figure 2d. In addition to the characteristic peaks of the starch itself, some new characteristic peaks appeared in the spectrum of St-g-PAM. There are sharp peaks at 1668 cm^{-1} and 1616 cm^{-1} that correspond to C=O stretching and N-H bending of the -CONH$_2$ group in acrylamide, respectively [28]. There is also an additional peak at 1454 cm^{-1} which is C-N bond stretching [29]. Thus, these results indicate that the acrylamide unit has been successfully grafted onto the starch backbone. Figure 2e shows the XRD patterns of the original starch and St-g-PAM. For the original starch, there are four obvious diffraction peaks at 2θ = 15.1°, 17.1°, 17.9° and 22.9°, but the intensity of the crystalline peaks of the modified grafted starch is obviously reduced. This shows that graft copolymerization increases the difficulty of neatly aligning the starch chain [30]. Thus, as the amount of monomer input increases, the grafting percentage gradually increases and the more PAM branches are introduced into the starch chain, the more significantly crystallinity decreases.

The TGA and DTG curves of starch and St-g-PAM samples at 30–650 °C under nitrogen are shown in Figure 2f,g. Starch involves two distinct mass loss regions, with an initial moisture mass loss at around 100 °C, which may be due to the separation of free water molecules present in starch. Moreover, there is a second mass loss zone between 250–330 °C, which is due to the breakdown of the glycosidic bonds in the starch chains. Finally, the grafted starch also has an additional region of mass loss (350–450 °C), which is due to the decomposition of the PAM grafted side chain portion of the starch chain, further indicating that the grafting of starch was successful [31].

3.3. Slurry and Film Properties

Hydration capacity is an important physicochemical property of slurry. Water solubility will lead to good permeability, flow and wettability, and good swelling will increase the adhesion of starch and the stability of slurry. Generally, the hydrophilicity of starch can be reflected by the degree of swelling of starch in water [32]. The water-soluble optical images of ACS and St-g-PAM samples at different temperatures are shown in Figure 3a. It is clearly observed that the pristine starch cannot be dissolved in the 55 °C water and is deposited in the bottom of the beaker. When the water is heated to 95 °C, starch is completely dissolved due to the water molecules connecting with the hydroxyl groups on the starch molecules through hydrogen bonding, resulting in the swelling and dissolution of the granules [33]. Compared with pristine starch, the solubility of St-g-PAM in water was significantly enhanced at the same temperature. In particular, St-g-PAM 3 and St-g-PAM 4 can form a uniform transparent solution in water at 55 °C due to the increase of amide groups, which are hydrophilic. The corresponding solubility is shown in Figure 3b. At 55 °C, the ACS is in a stratified state and the water solubility tends to 0. However, the solubility of the grafted starch is significantly increased. In addition, as the temperature rises, the solubility of ACS and St-g-PAM samples increased significantly, eventually reaching over 80%. Therefore the stronger the water solubility of the grafted starch, the more easily the particles are dispersed in water. Details of the corresponding water solubility are shown in Table S2. In addition, the swelling of ACS and St-g-PAM samples at different temperatures are shown in Figure S4. It is clearly observed that the swelling of all modified starches is better than that of the original starch (ACS) when the water temperature is below 85 °C. Swelling parameters in detail are listed in Table S3. Figure 3c shows the water solubilization time and moisture regain of ACS and St-g-PAM films. The branched chains grafted onto the starch chains greatly shortened the water solubilization time of the St-g-PAM films due to their hydrophilic properties and they were able to absorb moisture from the air, resulting in a significantly higher moisture regain of the St-g-PAM films than that of the ACS films. The transmittance of the slurry reflects the antiagglomeration performance of the slurry and directly reflects its dispersion stability. The solution of starch and modified starch dissolved in water at 90 °C was tested for transmittance, and the results are shown in Figure 3d. The transmittance increases gradually with the increase of PAM branches, and the transmittance of St-g-PAM 3 is comparable to that of ACS. The more hydrophilic branches are introduced, the more hydrophilic St-g-PAM is, and the more fully particles are dissolved. Optical photographs of the corresponding slurry films are shown in Figure S5.

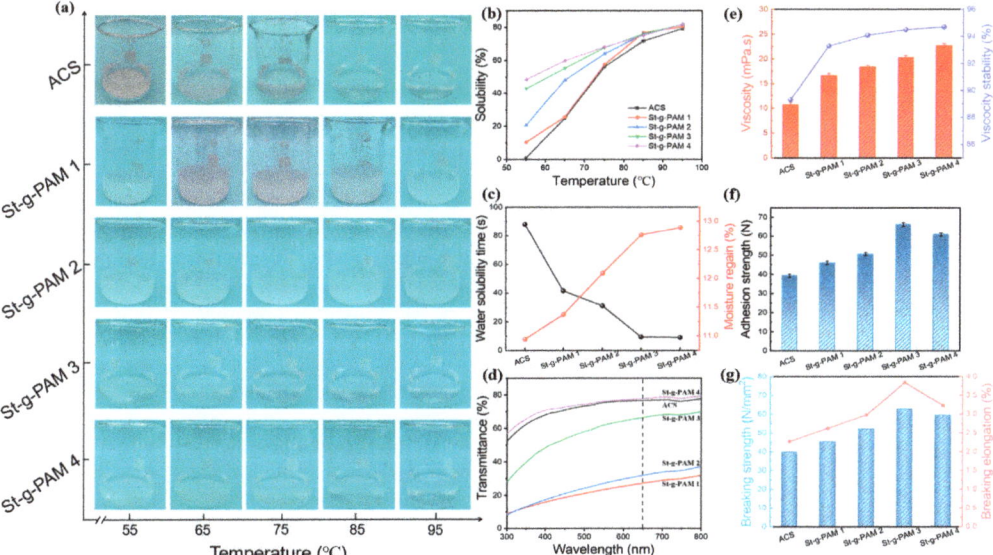

Figure 3. (**a**) Optical images depicting water solubility at different temperatures. (**b**) water solubility, (**c**) solubility time and moisture regain, (**d**) UV-visible transmittance of slurry ACS and St-g-PAM with different raw material ratios. (**e**) viscosity and stability, (**f**) adhesion, (**g**) breaking strength and elongation of slurry films ACS and St-g-PAM with different raw material ratios.

Viscosity and stability are important parameters for evaluating the performance of slurry. Graft polymerization has a significant effect on starch paste viscosity, as shown in Figure 3e, which indicates that the slurry viscosity is increasing with the increase of grafting percentage and the viscosity and stability of the modified starch paste has been greatly improved [34]. This may be attributed to the increase in molecular weight and the hydrophilicity of the grafted side chains, which improves the force between starch and water molecules, and increases the resistance to flow of the paste. Undoubtedly, the St-g-PAM slurry, with a stability greater than 90%, is fully capable of meeting the requirements for stability during the sizing process, thus ensuring the stability of sizing. The adhesion of the paste to the fibers helps to bond the fibers together to improve the strength and abrasion resistance of the sizing yarns, and to reduce the fluffiness or hairiness of the yarns, which ultimately improves the weavability of the sizing yarns [35]. The bonding properties of St-g-PAM with cotton fibers in different mass ratios were evaluated, as shown in Figure 3f. The results showed that the adhesion of St-g-PAM to cotton fibers was higher than that of ACS in all cases. With the increase of the ratio, the adhesion of cotton fibers increased from 39.3 N to 66.22 N and then decreased to 60.84 N, which indicated that the introduction of the PAM branch played a crucial role in improving the adhesion between starch and cotton fibers. When the mass ratio of starch and monomer was 1:2, the adhesion decreased slightly, indicating that the viscosity of the slurry was too high, which was unfavorable for yarn impregnation and draping. The cotton roving was impregnated in the slurry, then the slurry adhered to the surface of the roving, and finally a layer of adhesive was formed between the fibers. Figure 3g shows the effect of PAM branching on the tensile properties of the pulp films. The results show that the starch films prepared by graft copolymerization have superior elongation at break as well as tensile strength compared to the control starch ACS. From the figure, it can be seen that St-g-PAM 3 film has the highest elongation (3.84%) and the highest breaking strength (62.62 N · mm^{-2}), while the ACS film has the lowest elongation (2.27%) and the lowest breaking strength (39.86 N · mm^{-2}). This indicates that St-g-PAM films have superior properties compared to ACS films and are more suitable for warp sizing. The tensile strain of slurry films ACS and St-g-PAM with different raw material ratios are shown in Figure S6.

3.4. Sizing Yarn Morphology and Properties

In order to better evaluate the warp sizing performance of modified starch, we established a simple laboratory warp sizing device in a simulated real factory, as shown in Figure 4a. First, cotton denim yarns were sized through the sizing tank, then the residual sizing solution on the yarns was extruded using a squeezing roller, and finally the yarns were dried around the cylinder. The slurry concentration was 6%, the sizing temperature was 95 °C, the sizing speed was 30 m/min, and the drying temperature was controlled at 50 °C. The specific process parameters for pulping in the slurry tank are shown in Table S4. Optical microscope photographs and SEM images of the raw yarns and sizing yarns are shown in Figure 4b,c. It can be seen that the fibers of the raw yarn are loose and there is a lot of hairiness on the surface. After sizing, the slurry forms a film on the surface of the yarn, which makes the fibers stick together, thus reducing the hairiness of the surface of the yarn. Cross-sectional SEM images of the raw yarn and sizing yarn are shown in Figure S7, where it can be observed more intuitively that the fibers in the raw yarn are independent of each other, and the fibers in the sizing yarn are clustered together with smaller inter-fiber gaps. This shows that St-g-PAM is not only on the surface of the yarn but also penetrates into the yarn, which enhances the cohesion between the fibers and the ability to resist external forces, thus protecting the yarn and making it weavable.

The breaking elongation and abrasion resistance of sizing yarns before and after sizing are shown in Figure 4d. Compared with the pristine yarns, the warp yarn after sizing had higher wear resistance and lower elongation at break, which can be attributed to the superior adhesion of St-g-PAM slurry to cotton yarn, resulting in a more effective attachment of the slurry film onto the surface of cotton yarn. Consequently, this protective mechanism contributes to an improved wear resistance of cotton yarn. In addition, the breaking strength and moisture regain of the warp yarn after sizing was obviously improved, as shown in Figure S8. After sizing, the harmful hairiness on the surface of the warp was significantly reduced by more than 3 mm, and the moisture regain was consistent with the results of the sizing film, as shown in Figure 4e. The introduction of a hydrophilic PAM branch chain significantly enhanced the hydrophilicity and water dispersion of starch, thereby increasing the toughness of the starch gel layer while reducing break. During the sizing process, the slurry soaked into the interior of the yarn, which enhanced the holding force of the single fibers within the yarn and improved the strength of the yarn.

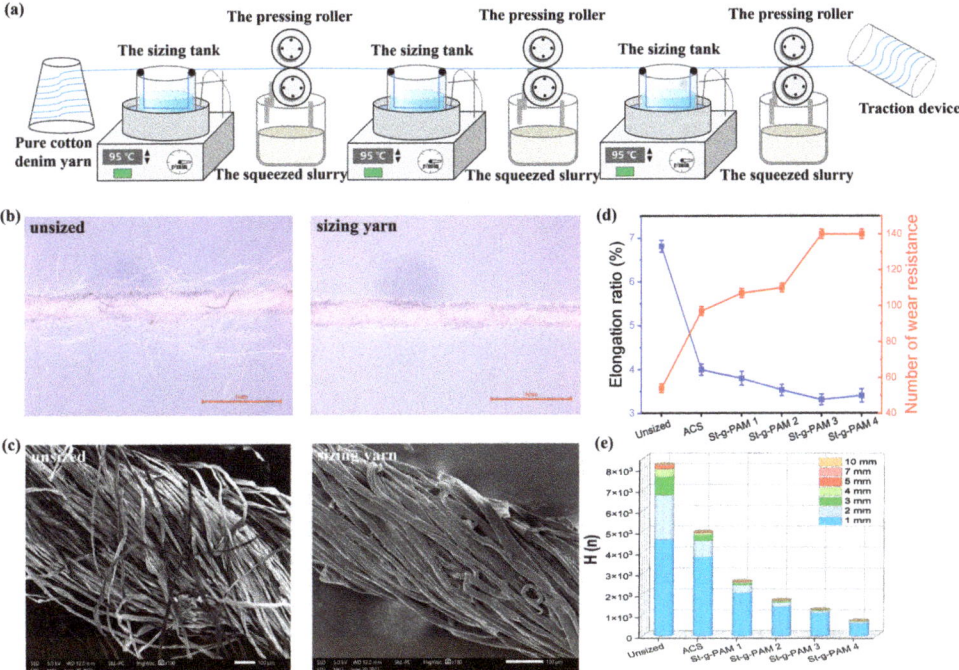

Figure 4. (**a**) Schematic diagram of sizing line. (**b**) Optical microscope images and (**c**) SEM images of unsized and sizing yarn. (**d**) Breaking elongation and abrasion resistance and (**e**) hair feathers of sizing yarns ACS and St-g-PAM with different raw material ratios.

3.5. Desizing Properties

In the textile process, the yarn frequently breaks, due to mechanical forces, so in the textile manufacturing process, warp sizing is used to improve the mechanical properties of the yarn, to better weave the fabric. However, prior to further dyeing procedures, it is imperative to eliminate the slurry adhered to the warp yarn in order to ensure optimal dyeing performance and fabric flexibility. Therefore, it is necessary to further evaluate the desizing performance of the sizing yarn. The desizing ratio, which measures the extent of desizing in relation to the upper sizing, is commonly employed as an indicator for evaluating the desizing ability of sizing yarns [36]. The optimal performance was observed with St-g-PAM 3 slurry, which has a starch to acrylamide mass ratio of 1:1.5, as discussed above. Consequently, we conducted further investigations on the desizing efficacy of the yarn treated with St-g-PAM 3 slurry. SEM images of the yarns after desizing the St-g-PAM 3 samples in water at different temperatures are shown in Figure 5a. It can be seen that the yarns were desized almost completely in water temperatures from 95 °C down to 75 °C, while there was a little residue of the slurry after desizing in water at 65 °C. Subsequently, we added 2–3 drops of a 10 g/L iodine aqueous solution onto the undesized yarn/fabric surface to facilitate a color reaction, thereby enabling a more visual assessment of any remaining starch content in the yarn. The undesized yarn/fabric exhibits a dark blue color upon encountering iodine, as depicted in Figure 5b. Subsequent desizing in water at 65 °C resulted in a lighter coloration, with only a faint shade of blue remaining, indicating the presence of residual modified starch on the surface of the yarn/fabric. This indicates that grafted starch can be desized in water temperatures in the 65–75 °C range, which reduces the energy consumption needed for high temperature desizing.

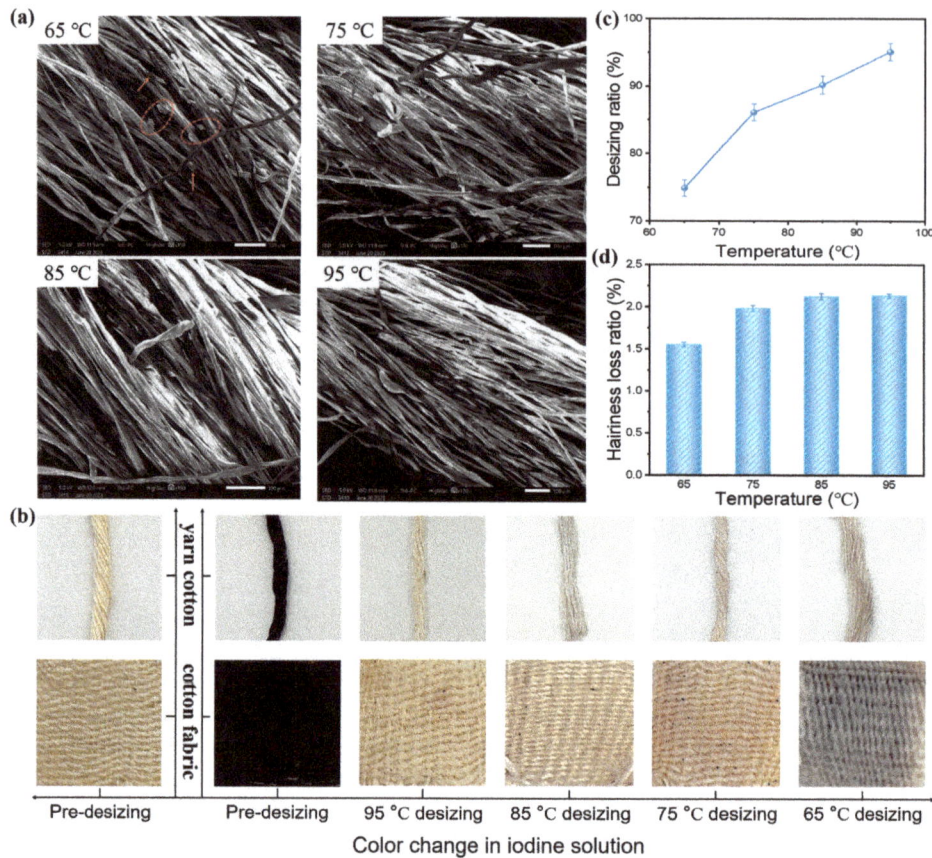

Figure 5. (a) SEM micrographs and (b) hairiness loss ratio after desizing at different temperatures. (c) Desizing ratio at different temperatures. (d) Color change in iodine solution before and after desizing.

Furthermore, we conducted calculations on the desizing ratio of St-g-PAM 3 sizing in water at various temperatures, as illustrated in Figure 5c. The results indicate that the desizing ratio of the sizing increases proportionally with temperature elevation due to the degradation of the starch molecule structure and subsequent enhancement of its solubility, which is consistent with the results in Figure 3a above. Generally, in the desizing operation, more than 80% of starch should be removed from the sizing yarn, leaving residual sizing after desizing of less than 1% of the fabric weight, in order not to affect the subsequent process [37]. Therefore, when the desizing temperature is controlled at about 70 °C, it can meet the desizing requirements. In the desizing process, the heavy loss ratio of hairiness will cause a decrease in the mechanical properties of the yarn. Therefore, we evaluated the hairiness loss ratio of yarns desizing at different temperatures, as shown in Figure 5d. When the desizing temperature ranged from 65 °C to 95 °C, the yarn hairiness loss ratio was 1.55% (65 °C), 1.98% (75 °C), 2.12% (85 °C) and 2.13% (95 °C), respectively. It is obvious that the yarn hair loss ratio increases with the increase of the desizing temperature. However, as a whole, between the temperatures of 65 °C and 95 °C, the yarn hair loss ratio was relatively small, at less than 2.2%, which meets the requirements of yarn sizing [38].

Finally, we investigated the biochemical degradability by testing the BOD_5/COD_{cr} values of the desizing wastewater of the St-g-PAM 3 prepared in this study and two commercially available PVA slurries. It was generally accepted that $BOD_5/COD_{cr} > 0.45$

indicated excellent biochemical degradability, and $BOD_5/COD_{cr} < 0.25$ indicated poor biochemical treatment. As can be seen from Table 1, the B/D values of the two PVA slurries were much less than 0.25, which indicated that they would be difficult to biochemically treat, while that of the grafted starch slurry had a B/D of = 0.476, which revealed an excellent biochemical degradability, and thus, St-g-PAM 3 has significant advantages and meets the requirements of the development of the modern textile industry. In terms of energy saving, the sizing process included: slurry boiling paste, pipeline slurry transfer, slurry tank soaking, slurry roller pressing (three cycles), drying, and manufacture by weaving. By adopting wet sizing technology, using squeezing rollers to control the yarn's roll residue after dyeing and washing at about 60%, the yarn's drying step before sizing is omitted. This changes the traditional sizing that needed to be carried out in the drying process before sizing, reducing by 30% the amount of steam in the process of drying before sizing, and therefore reducing the cost of production. The new technology is now the subject of a patent application. This method uses low temperatures (60 °C) for washing and desizing, requires no chemical additives, and saves energy consumption.

Table 1. BOD_5 and COD_{cr} values of slurries.

	St-g-PAM	1788PVA	1799PVA
BOD_5 (mg/L)	39,400	1630	800
COD_{cr} (mg/L)	82,773	181,000	182,000
BOD_5/COD_{cr}	0.476	0.009	0.004

4. Conclusions

St-g-PAM pastes with different grafting percentages were prepared using different mass ratios of starch and monomer. The results showed that the graft copolymerization of starch with AM could overcome the drawbacks of starch such as brittleness in warp sizing and improve its end use properties. Basic characterization, such as infrared spectroscopy and EDS, confirmed that the successful introduction of PAM branches increased the hydrophilicity, anticoagulation and viscosity stability of starch, thus effectively improving the adhesion properties between starch and cotton fibers. The grafted PAM side chains shortened the rupture time of the pulp film in hot water and enhanced the tensile properties of the film as well as the abrasion resistance of the sizing yarn. These results indicate that the introduced PAM branched chains can reduce the brittleness of starch film and increase the toughness to protect the yarn and meet the weaving requirements. In addition, a comprehensive analysis of the performance of the slurry film and sizing yarns determined that St-g-PAM 3 slurry is optimal when the mass ratio of starch to acrylamide is 1:1.5, while maintaining a good desizing effect in hot water, and the above suggests that St-g-PAM slurry is expected to be used as a new type of environmentally friendly slurry for sizing cotton yarns.

Supplementary Materials: The following supporting information can be downloaded at: https://www.mdpi.com/article/10.3390/polym16020182/s1, Figure S1: The Synthesis process of grafted starch slurry, Figure S2: SEM images of St-g-PAM 1, St-g-PAM 2 and St-g-PAM 4, Figure S3: EDS spectrogram of St-g-PAM 1, St-g-PAM 2 and St-g-PAM 4, Figure S4: Swelling power of slurry ACS and St-g-PAM with different raw material ratios, Figure S5: Optical photographs of the appearance of the slurry films ACS and St-g-PAM with different raw material ratios, Figure S6: Tensile strain of slurry films ACS and St-g-PAM with different raw material ratios, Figure S7: SEM cross-section images of unsized and sizing yarn, Figure S8: Breaking strength and moisture regain of sizing yarns ACS and St-g-PAM with different raw material ratios. Table S1: Specific synthetic details of graft copolymers, Table S2: Water solubility of St-g-PAM with different ratios at different temperatures, Table S3: Swelling power of St-g-PAM with different ratios at different temperatures, Table S4: Basic properties of the slurry.

Author Contributions: Conceptualization, Y.Z. and J.L.; methodology, Y.Z.; software, Z.W.; validation, Y.Z., F.G. and Z.L.; formal analysis, Z.L.; investigation, C.Y.; resources, Z.W.; data curation,

Y.Z.; writing—original draft preparation, Y.Z.; writing—review and editing, Y.Z.; visualization, Y.Z.; supervision, Z.L.; project administration, C.Y.; funding acquisition, C.Y. All authors have read and agreed to the published version of the manuscript.

Funding: The authors are grateful for the Key Research and Development Program of Hubei Province, China (Grant 2022BAA096).

Institutional Review Board Statement: Not applicable.

Data Availability Statement: The data presented in this study are available upon request from the corresponding author.

Acknowledgments: Z.H.L. acknowledges the support from the Ministry of Education Key Laboratory for the Green Preparation and Application Functional Materials, Hubei Key Laboratory of Polymer Materials, Hubei University.

Conflicts of Interest: The authors declare that they have no known competing financial interests or personal relationships that could have appeared to influence the work reported in this paper.

References

1. Liu, F.; Zhu, Z.; Xu, Z.; Zhang, X. Desizability of the Grafted Starches Used as Warp Sizing agents. *Starch-Starke* **2018**, *70*, 3–4. [CrossRef]
2. Zhu, Z.; Cao, S. Modifications to Improve the Adhesion of Crosslinked Starch Sizes to Fiber Substrates. *Text. Res. J.* **2004**, *74*, 253–258. [CrossRef]
3. Bismark, S.; Xun, Z.; Zhifeng, Z.; Charles, F.; Williams, B.; Benjamin, A.; Ebenezer, H.K. Phosphorylation and Octenylsuccinylation of Acid-Thinned Starch for Enhancing Adhesion on Cotton/Polyester Blend Fibers at Varied Temperature Sizing. *Starch-Starke* **2018**, *71*, 1800055. [CrossRef]
4. Yu, L.; Dean, K.; Li, L. Comprehensive review on single and dual modification of starch: Methods, properties and applications. *Int. J. Biol. Macromol. Sci.* **2023**, *253*, 126952.
5. Tao, K.; Li, C.; Yu, W.; Gilbert, R.G.; Li, E. How amylose molecular fine structure of rice starch affects functional properties. *Carbohydr. Polym.* **2018**, *204*, 24–31. [CrossRef] [PubMed]
6. Li, W.; Zhang, Z.; Wu, L.; Zhu, Z.; Ni, Q.; Xu, Z.; Wu, J. Cross-linking/sulfonation to improve paste stability, adhesion and film properties of corn starch for warp sizing. *Int. J. Adhes. Adhes.* **2021**, *104*, 102720. [CrossRef]
7. Li, W.; Zhu, Z. Electroneutral maize starch by quaterization and sulfosuccination forstrong adhesion-to-viscose fibers and easy removal. *J. Adhes.* **2015**, *92*, 257–272. [CrossRef]
8. Castanha, N.; Lima, D.C.; Junior, M.D.M.; Campanella, O.H.; Augusto, P.E.D. Combining ozone and ultrasound technologies to modify maize starch. *Int. J. Biol. Macromol.* **2019**, *139*, 63–74. [CrossRef]
9. Mostafa, K.M.; El-Sanabary, A.A. Harnessing of novel tailored modified pregelled starch-derived products in sizing of cotton textiles. *Adv. Polym. Technol.* **2011**, *31*, 52–62. [CrossRef]
10. Bao, X.; Yu, L.; Shen, S.; Simon, G.P.; Liu, H.; Chen, L. How rheological behaviors of concentrated starch affect graft copolymerization of acrylamide and resultant hydrogel. *Carbohydr. Polym.* **2019**, *219*, 395–404. [CrossRef]
11. Pourmahdi, M.; Abdollahi, M.; Nasiri, A. Effect of lignin source and initiation conditions on graft copolymerization of lignin with acrylamide and performance of graft copolymer as additive in water-based drilling fluid. *J. Pet. Sci. Eng.* **2023**, *220*, 111253. [CrossRef]
12. Bolat, K.; Hasanoğlu, A.; Seçer, A. Use of modified corn starches as environmental and cost-friendly alternatives of PVA in sizing applications. *J. Text. Inst.* **2020**, *112*, 1688–1699. [CrossRef]
13. Jiang, T.; Chen, F.; Duan, Q.; Bao, X.; Jiang, S.; Liu, H.; Chen, L.; Yu, L. Designing and application of reactive extrusion with twice initiations for graft copolymerization of acrylamide on starch. *Eur. Polym. J.* **2022**, *165*, 111008. [CrossRef]
14. Siyamak, S.; Luckman, P.; Laycock, B. Rapid and solvent-free synthesis of pH-responsive graft-copolymers based on wheat starch and their properties as potential ammonium sorbents. *Int. J. Biol. Macromol.* **2020**, *149*, 477–486. [CrossRef] [PubMed]
15. Zhu, Z.; Zhu, Z. Adhesion of starch-g-poly(2-acryloyloxyethyl trimethyl ammonium chloride) to cotton and polyester fibers. *Starch-Starke* **2014**, *66*, 566–575. [CrossRef]
16. Zha, X.; Sadi, S.; Yang, Y.; Luo, T.; Huang, N. Introduction of poly(acrylic acid) branch onto acetate starch for polyester warp sizing. *J. Text. Inst.* **2020**, *112*, 273–285. [CrossRef]
17. Zhu, Z.; Wang, M.; Li, W. Starch maleation and sulfosuccinylation to alleviate the intrinsic drawback of brittleness of cornstarch film for warp sizing. *Fibers Polym.* **2015**, *16*, 1890–1897. [CrossRef]
18. Djordjevic, S.; Kovacevic, S.; Nikolic, L.; Miljkovic, M.; Djordjevic, D. Cotton Yarn Sizing by Acrylamide Grafted Starch Copolymer. *J. Nat. Fibers* **2014**, *11*, 212–224. [CrossRef]
19. Obadi, M.; Qi, Y.; Xu, B. High-amylose maize starch: Structure, properties, modifications and industrial applications. *Carbohydr. Polym.* **2023**, *299*, 120185. [CrossRef]

20. Zhang, X.; Baek, N.-W.; Lou, J.; Xu, J.; Yuan, J.; Fan, X. Effects of exogenous proteins on enzyme desizing of starch and its mechanism. *Int. J. Biol. Macromol.* **2022**, *218*, 375–383. [CrossRef]
21. Tu, Y.; Shao, G.; Zhang, W.; Chen, J.; Qu, Y.; Zhang, F.; Tian, S.; Zhou, Z.; Ren, Z. The degradation of printing and dyeing wastewater by manganese-based catalysts. *Sci. Total. Environ.* **2022**, *828*, 154390. [CrossRef]
22. Zhou, H.; Zhou, L.; Ma, K. Microfiber from textile dyeing and printing wastewater of a typical industrial park in China: Occurrence, removal and release. *Sci. Total. Environ.* **2020**, *739*, 140329. [CrossRef] [PubMed]
23. Shahzad, H.M.A.; Khan, S.J.; Zeshan; Jamal, Y.; Habib, Z. Evaluating the performance of anaerobic moving bed bioreactor and upflow anaerobic hybrid reactor for treating textile desizing wastewater. *Biochem. Eng. J.* **2021**, *174*, 108123. [CrossRef]
24. Liu, S.-S.; You, W.-D.; Chen, C.-E.; Wang, X.-Y.; Yang, B.; Ying, G.-G. Occurrence, fate and ecological risks of 90 typical emerging contaminants in full-scale textile wastewater treatment plants from a large industrial park in Guangxi, Southwest China. *J. Hazard. Mater.* **2023**, *449*, 131048. [CrossRef] [PubMed]
25. Zhu, Z.; Chen, P. Carbamoyl ethylation of starch for enhancing the adhesion capacity to fibers. *J. Appl. Polym. Sci.* **2007**, *106*, 2763–2768. [CrossRef]
26. Li, W.; Xu, Z.; Wang, Z.; Liu, X.; Li, C.; Ruan, F. Double etherification of corn starch to improve its adhesion to cotton and polyester fibers. *Int. J. Adhes. Adhes.* **2018**, *84*, 101–107. [CrossRef]
27. Li, Y.; Yang, J. Quantifying the microstructure evolution and phase assemblage of solid waste as construction materials by EDS image analysis: A case study of carbonated basic oxygen furnace slag aggregates. *Constr. Build. Mater.* **2023**, *393*, 131892. [CrossRef]
28. Nguyen, L.N.; Vu, H.P.; Fu, Q.; Johir, A.H.; Ibrahim, I.; Mofijur, M.; Labeeuw, L.; Pernice, M.; Ralph, P.J.; Nghiem, L.D. Synthesis and evaluation of cationic polyacrylamide and polyacrylate flocculants for harvesting freshwater and marine microalgae. *Chem. Eng. J.* **2021**, *433*, 133623. [CrossRef]
29. Wang, L.; Zhang, X.; Xu, J.; Wang, Q.; Fan, X. How starch-g-poly(acrylamide) molecular structure effect sizing properties. *Int. J. Biol. Macromol.* **2019**, *144*, 403–409. [CrossRef]
30. Shen, Y.; Yao, Y.; Wang, Z.; Wu, H. Hydroxypropylation reduces gelatinization temperature of corn starch for textile sizing. *Cellulose* **2021**, *28*, 5123–5134. [CrossRef]
31. Mishra, S.; Mukul, A.; Sen, G.; Jha, U. Microwave assisted synthesis of polyacrylamide grafted starch (St-g-PAM) and its applicability as flocculant for water treatment. *Int. J. Biol. Macromol.* **2011**, *48*, 106–111. [CrossRef] [PubMed]
32. Chen, Y.; Duan, Q.; Zhu, J.; Liu, H.; Chen, L.; Yu, L. Anchor and bridge functions of APTES layer on interface between hydrophilic starch films and hydrophobic soyabean oil coating. *Carbohydr. Polym.* **2021**, *272*, 118450. [CrossRef] [PubMed]
33. Zhang, L.; Xiong, T.; Wang, X.-F.; Chen, D.-L.; He, X.-D.; Zhang, C.; Wu, C.; Li, Q.; Ding, X.; Qian, J.-Y. Pickering emulsifiers based on enzymatically modified quinoa starches: Preparation, microstructures, hydrophilic property and emulsifying property. *Int. J. Biol. Macromol.* **2021**, *190*, 130–140. [CrossRef] [PubMed]
34. Lee, H.; Kim, H.-S. Pasting and paste properties of waxy rice starch as affected by hydroxypropyl methylcellulose and its viscosity. *Int. J. Biol. Macromol.* **2019**, *153*, 1202–1210. [CrossRef] [PubMed]
35. Li, W.; Yu, Y.; Wu, Y.; Liu, Q. Preparation, characterization of feather protein-g-poly(sodium allyl sulfonate) and its application as a low-temperature adhesive to cotton and viscose fibers for warp sizing. *Eur. Polym. J.* **2020**, *136*, 109945. [CrossRef]
36. Li, W.; Zhang, Z.; Wu, L.; Liu, Q.; Cheng, X.; Xu, Z. Investigating the relationship between structure of itaconylated starch and its sizing properties: Viscosity stability, adhesion and film properties for wool warp sizing. *Int. J. Biol. Macromol.* **2021**, *181*, 291–300. [CrossRef]
37. Li, W.; Wu, L.; Zhang, Z.; Ke, H.; Liu, Q.; Zhu, Z.; Xu, J.; Wei, A.; Cheng, X. Introduction of poly(2-acrylamide-2-methylpropanesulfonic acid) branches into starch molecules for improving its paste stability, adhesion and desizability. *Int. J. Adhes. Adhes.* **2021**, *110*, 102939. [CrossRef]
38. Li, W.; Wu, L.; Zhu, Z.; Zhang, Z.; Liu, Q.; Lu, Y.; Ke, H. Incorporation of poly(sodium allyl sulfonate) branches on corn starch chains for enhancing its sizing properties: Viscosity stability, adhesion, film properties and desizability. *Int. J. Biol. Macromol.* **2020**, *166*, 1460–1470. [CrossRef]

Disclaimer/Publisher's Note: The statements, opinions and data contained in all publications are solely those of the individual author(s) and contributor(s) and not of MDPI and/or the editor(s). MDPI and/or the editor(s) disclaim responsibility for any injury to people or property resulting from any ideas, methods, instructions or products referred to in the content.

Article

Rigidity with Flexibility: Porous Triptycene Networks for Enhancing Methane Storage

Fei Guo [1,†], Hui Ma [2,†], Bin-Bin Yang [2,†], Zhen Wang [1,2,*], Xiang-Gao Meng [3,*], Jian-Hua Bu [4] and Chun Zhang [2]

1. National Engineering Laboratory for Advanced Yarn and Fabric Formation and Clean Production, Technology Institute, Wuhan Textile University, Wuhan 430200, China; guofeifei0806@163.com
2. College of Life Science and Technology, National Engineering Research Center for Nanomedicine, Huazhong University of Science and Technology, Wuhan 430074, China; dongji0828@126.com (H.M.); 17371266359@163.com (B.-B.Y.); chunzhang@hust.edu.cn (C.Z.)
3. School of Chemistry, Central China Normal University, Wuhan 430079, China
4. Xi'an Modern Chemistry Research Institute, Xi'an 710065, China; bujianhua@163.com
* Correspondence: wz@wtu.edu.cn (Z.W.); xianggao_meng@126.com (X.-G.M.)
† These authors contributed equally to this work.

Citation: Guo, F.; Ma, H.; Yang, B.-B.; Wang, Z.; Meng, X.-G.; Bu, J.-H.; Zhang, C. Rigidity with Flexibility: Porous Triptycene Networks for Enhancing Methane Storage. *Polymers* 2024, *16*, 156. https://doi.org/10.3390/polym16010156

Academic Editor: Alessio Fuoco

Received: 28 September 2023
Revised: 20 November 2023
Accepted: 24 November 2023
Published: 4 January 2024

Copyright: © 2024 by the authors. Licensee MDPI, Basel, Switzerland. This article is an open access article distributed under the terms and conditions of the Creative Commons Attribution (CC BY) license (https://creativecommons.org/licenses/by/4.0/).

Abstract: In the pursuit of advancing materials for methane storage, a critical consideration arises given the prominence of natural gas (NG) as a clean transportation fuel, which holds substantial potential for alleviating the strain on both energy resources and the environment in the forthcoming decade. In this context, a novel approach is undertaken, employing the rigid triptycene as a foundational building block. This strategy is coupled with the incorporation of dichloromethane and 1,3-dichloropropane, serving as rigid and flexible linkers, respectively. This combination not only enables cost-effective fabrication but also expedites the creation of two distinct triptycene-based hypercrosslinked polymers (HCPs), identified as PTN-70 and PTN-71. Surprisingly, despite PTN-71 manifesting an inferior Brunauer–Emmett–Teller (BET) surface area when compared to the rigidly linked PTN-70, it showcases remarkably enhanced methane adsorption capabilities, particularly under high-pressure conditions. At a temperature of 275 K and a pressure of 95 bars, PTN-71 demonstrates an impressive methane adsorption capacity of 329 cm^3 g^{-1}. This exceptional performance is attributed to the unique flexible network structure of PTN-71, which exhibits a pronounced swelling response when subjected to elevated pressure conditions, thus elucidating its superior methane adsorption characteristics. The development of these advanced materials not only signifies a significant stride in the realm of methane storage but also underscores the importance of tailoring the structural attributes of hypercrosslinked polymers for optimized gas adsorption performance.

Keywords: triptycene; hypercrosslinked polymers; flexibility; methane storage

1. Introduction

According to the European Union Commission report stated in 2004, transportation was one of the main sectors contributing to energy consumption and depletion, of which harmful emissions resulted in global warming, climatic deterioration, and air pollution. As a response to these environmental challenges, there has been a growing interest in using natural gas, primarily composed of methane (>95%), as a cleaner alternative to conventional energy sources like oil-derived products [1–3]. However, a critical hurdle in realizing the potential of natural gas (NG) as a clean fuel lies in the efficient storage of methane, particularly for applications in transportation [4]. This challenge has spurred research into porous adsorbents with high surface areas and proper pore volumes, which have the potential to enable efficient methane storage. Among these porous materials are molecular cages [5,6], porous organic polymers (POPs) [7–10], and metal-organic frameworks (MOFs) [11–14]. Among them, hypercrosslinked polymers (HCPs), formed by a simpler and scalable procedure with irreversible condensation of Friedel–Crafts alkylation or Scholl reaction, possess generally outstanding physicochemical stability compared with MOFs and COFs [15–17].

Triptycene and its derivatives, renowned for their captivating three-dimensional and rigid molecular structures, have emerged as focal points of extensive research across diverse domains within chemistry and materials science. Their distinctive structural characteristics, characterized by three fused aromatic rings forming a compact triangular arrangement, bestow upon them a unique and intriguing nature. These compounds are particularly valued for their pronounced steric hindrance, which allows precise control of chemical reactivity, a trait of paramount importance in selective reactions. Furthermore, triptycene's ability to exhibit chirality makes it invaluable in asymmetric synthesis and as chiral catalysts. Their three-dimensional architecture offers exceptional versatility, permitting the attachment of various functional groups at different positions, thus serving as a versatile building block for the creation of novel molecules and advanced materials. Notably, the rigid framework of triptycene and its derivatives finds application in materials science, giving rise to materials with remarkable mechanical and electronic properties, thereby contributing to the development of high-performance polymers and organic electronic devices [18–21]. However, when used as building blocks for creating porous materials, triptycene-based structures often exhibit fixed and ungovernable porous properties, limiting their adaptability for specific applications. Recently, connecting rigid monomers with flexible linkers showed distinguished methane adsorption under high pressure, which held promise as an efficient storage method for NG in motor vehicles. A noteworthy concept introduced by Jeffrey R. Long and colleagues involves the observation of "gate-opening" behavior in flexible metal-organic frameworks (MOFs). This behavior is characterized by distinctive "S-shaped" or "stepped" adsorption isotherms, which have proven advantageous for methane storage applications [22]. Recently, Cafer T. Yavuz et al. constructed a carbon-carbon bonded HCP (COP-150) by benzene and 1,2-dichloroethane as both solvent and flexible linker, achieving the targets the US Department of Energy (DOE) set with high deliverable methane capacity (0.625 g g^{-1} and 294 L L^{-1}), which benefited from the flexible mechanism [23]. The flexibility affords the transformation of HCP networks depending on gaseous species from a nonporous or low-porous system to a porous system, which is highlighted for gas adsorption [24–26].

The inherent challenge of achieving a delicate balance between maintaining the structural integrity of a network while incorporating highly flexible groups remains a significant hurdle in material science. The total collapse of a network often arises when the structural units predominantly consist of highly flexible groups. To address this challenge, an effective strategy involves a complementary approach that combines a rigid, contorted skeleton with flexible chain segments. This approach has proven to be instrumental in overcoming the structural collapse associated with highly flexible frameworks [27–29]. Previous reports from our group highlighted the successful synthesis of diverse hypercrosslinked polymers (HCPs) based on triptycene and its derivatives as monomers, characterized by their three-dimensional rigid paddle wheel shape. These HCPs demonstrated excellent performance in adsorbing organic dyes and exhibited high CO_2 uptake owing to their superior surface area and porosity [30–34]. Building on this foundation, the present work aimed to explore the distinctiveness of methane adsorption behavior between rigid and flexible linkers in the networks of HCPs (Figure 1). A series of triptycene-based HCPs were synthesized in different solvents, namely dichloromethane and 1,3-dichloropropane, to systematically investigate the impact of linker flexibility on methane adsorption. The synthesis process was designed to be simultaneously convenient and economically feasible, occurring at room temperature without the need for expensive metal catalysts or intricate purification steps. Intriguingly, PTN-71, characterized by a significantly lower surface area (S_{BET} = 574 m^2 g^{-1}) compared to PTN-70 (S_{BET} = 1873 m^2 g^{-1}), exhibited a notably higher methane adsorption capacity (329 m^3 g^{-1}). This unexpected result underscores the crucial role of the flexible mechanism within PTN-71, activated under pressure ranging from 0.05 to 100 bars, in achieving enhanced methane adsorption despite the surface area disparity. The systematic exploration of different solvents and the tailored synthesis approach contribute to unraveling the intricate interplay between structural design, flexibility, and gas adsorption capabilities in hypercross-linked polymers. The outcomes of this study not only advance our understanding of material science but also offer insights into the development of cost-effective and efficient strategies for methane storage, opening avenues for sustainable energy applications.

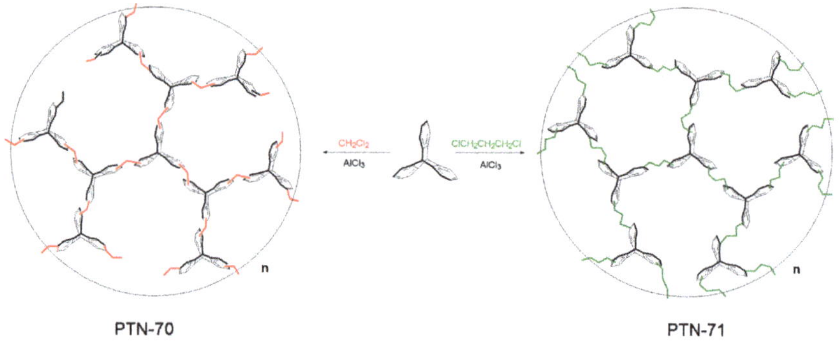

Figure 1. Synthesis of PTN-70 and PTN-71.

2. Materials and Methods

2.1. Materials and Instrumentation

All reagents utilized in this study were procured from commercial suppliers and employed without additional purification. The 13C CP/MAS NMR spectra were acquired with a contact time of 2 ms (ramp 100) and a pulse delay of 3 s.

The X-ray intensity data for PTN-70 and PTN-71 were collected using a standard Bruker SMART-1000 CCD Area Detector System, equipped with a normal-focus Cu-target X-ray tube (Bruker, Germany). Fourier-transform infrared (FT-IR) spectra were recorded on a Bruker model VERTEX 70 infrared spectrometer.

Thermogravimetric analysis (TGA) measurements were conducted on a PerkinElmer model Pyrisl TGA (PerkinElmer Instruments, Shanghai, China) under a nitrogen atmosphere, heating to 800 °C at a rate of 10 °C min^{-1}. Surface areas and pore size distributions were determined through nitrogen adsorption and desorption at 77 K, utilizing a Micromeritics ASAP 2020 volumetric adsorption analyzer (PerkinElmer Instruments, Shanghai, China) under a nitrogen atmosphere, heating to 800 °C at a rate of 10 °C min^{-1}. Surface areas and pore size distributions were determined through nitrogen adsorption and desorption at 77 K, utilizing a Micromeritics ASAP 2020 volumetric adsorption analyzer (Micromeritics Instrument Crop, Shanghai, China).

The specific rotation of both monomers and polymers was assessed using Autopol IV. Field-emission scanning electron microscopy (FE-SEM) measurements were carried out on a Tescan VEGA 3 SBH field-emission scanning electron microscope (Tescan, Shanghai, China). Transmission electron microscopy (TEM) studies were performed using a Tecnai G220 electron microscope (FEI, America).

2.2. Synthesis of PTN-70 and PTN-71

2.2.1. Synthesis of PTN-70

In this synthetic procedure, triptycene (3.18 g, 12.5 mmol, 1 equiv) and AlCl$_3$ (6.72 g, 50 mmol, 4 equiv) were meticulously dissolved in 80 mL of dichloromethane within an argon atmosphere. The resultant solution mixture underwent stirring at room temperature for a duration of 3 days. Subsequently, to bring the reaction to completion, 150 mL of methanol was introduced into the mixture. The ensuing suspension was then subjected to filtration, followed by thorough washing with methanol and dichloromethane, each performed three times. To further refine the product, Soxhlet extraction with methanol and dichloromethane was carried out over a period of 3 days. Finally, the resulting product was subjected to vacuum drying for 2 days, ultimately yielding 3.07 g (89%).

2.2.2. Synthesis of PTN-71

In this synthetic procedure, triptycene (3.18 g, 12.5 mmol, 1 equiv) and AlCl$_3$ (6.72 g, 50 mmol, 4 equiv) were meticulously dissolved in 80 mL of 1,3-dichloropropane within an argon atmosphere. The resultant solution mixture underwent stirring at room temperature for a duration of 3 days. Subsequently, to bring the reaction to completion, 150 mL of methanol was introduced into the mixture. The ensuing suspension was then subjected to filtration, followed by thorough washing with methanol and dichloromethane, each performed three times. To further refine the product, Soxhlet extraction with methanol and dichloromethane was carried out over a period of 3 days. Finally, the resulting product was subjected to vacuum drying for 2 days, ultimately yielding 2.80 g (66%).

3. Results

Triptycene, an exceptional and inherently rigid aromatic compound, served as a pivotal precursor in the intricate synthesis of two distinct hypercrosslinked polymers, designated as PTN-70 and PTN-71. This synthetic endeavor unfolded within the realm of organic chlorinated solvents, with dichloromethane and 1,3-dichloropropane assuming dual roles as both reactants and linkers in the formation of these polymers. The catalytic influence of aluminum chloride emerged as a critical factor, orchestrating the polymerization reactions with finesse, and all of this was orchestrated at room temperature. The nuanced structural characteristics of PTN-70 and PTN-71 demanded a meticulous exploration, prompting a comprehensive array of characterizations. This included the application of sophisticated techniques such as Fourier-transform infrared (FT-IR) spectroscopy and solid-state nuclear magnetic resonance (^{13}C CP/MAS NMR) to unravel the intricacies of their molecular architectures. As shown in Figure S1, the FT-IR spectra of PTN-70 and PTN-71, when compared to that of the triptycene, revealed a series of distinctive features. Notably, both polymers exhibited prominent peaks in the range of 600–900 cm^{-1}, corresponding to the aromatic C-H bending vibrations, indicative of the presence of aromatic moieties in their structures. Additionally, aromatic C-C stretching peaks were observed at approximately 1500 cm^{-1}, further confirming the retention of the aromatic character. The aromatic C-H stretching vibrations, occurring above 3000 cm^{-1}, signified the continued existence of aromatic rings within the polymer structures. However, the most conspicuous differences were observed in the regions associated with methylene groups. In both PTN-70 and PTN-71, symmetric and asymmetric C-H stretching bands of methylene were clearly evident at 2856 and 2927 cm^{-1}, respectively, indicative of the presence of these methylene units in the polymer backbone. Furthermore, the -CH-bending peaks of methylene were observed at 1380 cm^{-1}, further reinforcing the conclusion that methylene groups played a significant role in the polymerization process, likely serving as integral components of the linkers that connected the triptycene units. The spectroscopic results obtained from various techniques collectively provide compelling evidence affirming the successful synthesis of PTN-70 and PTN-71, emphasizing the preservation of the triptycene core and the incorporation of methylene linkers into their respective structures. In the domain of ^{13}C CP/MAS NMR spectroscopy (Figure S2), a rich array of signals emerges; particularly notable are the aromatic carbon resonances at 127, 132, and 141 ppm, along with the distinctive signal arising from the methylidyne bridge carbon in triptycene, approximately 80 ppm. The signals observed in the 10–55 ppm range have been confidently assigned to methylene groups, with their broad signal range attributed to varying degrees of shielding effects, resulting in distinct chemical environments for these methylene moieties. Notably, in comparison to PTN-70, the heightened flexibility introduced by 1,3-dichloropropane in PTN-71 facilitates the creation of a higher packing density within the material. Consequently, this enhanced packing density contributes to stronger shielding effects, leading to a shift of the methylene signals in PTN-71 towards the high-field region. Furthermore, an intriguing signal beyond

170 ppm emerges, attributed to the presence of CO_2 infiltrating the porous networks when exposed to atmospheric air. These comprehensive spectroscopic findings not only shed light on the distinct structural attributes and chemical compositions of the synthesized materials but also underscore the potential for varied applications of PTN-70 and PTN-71 in the realm of porous materials and gas adsorption.

In a dedicated pursuit of unraveling the intricate morphological characteristics inherent in PTN-70 and PTN-71, a sophisticated approach was adopted, leveraging advanced imaging techniques, specifically field-emission scanning electron microscopy (FE-SEM) and transmission electron microscopy (TEM). The outcome of these meticulous analyses provided a wealth of insights into the structural attributes of these hypercrosslinked polymers, unveiling visually intriguing features that characterize both PTN-70 and PTN-71. The captivating images presented in Figure 2 and Figure S3 offered a glimpse into the world of amorphous, rough spherical particles, each adorned with irregular structures. What emerges as particularly striking in this exploration is the marked distinction observed in the morphology between PTN-70 and PTN-71. PTN-71, in particular, commands attention due to the notably higher packing density of its amorphous, rough spherical particles in comparison to its counterpart, PTN-70. This dissimilarity in packing density, evident in the vivid imagery captured by the advanced microscopy techniques, was further substantiated by the outcomes of powder X-ray diffraction (PXRD) analyses, as eloquently showcased in Figure S4. The PXRD analyses laid bare broad peaks, a characteristic feature that unequivocally signifies the noncrystalline nature inherent in both PTN-70 and PTN-71.

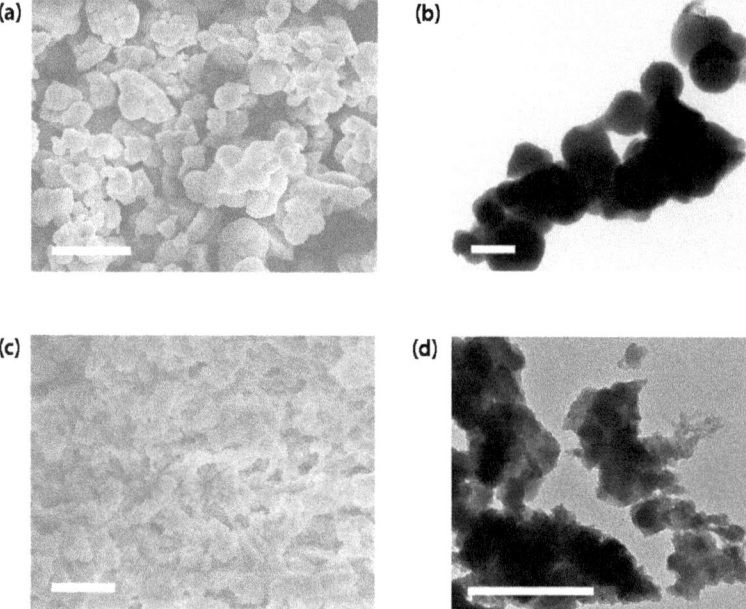

Figure 2. SEM and TEM images of PTN-70 (**a**,**b**) and PTN-71 (**c**,**d**). Scale bar: 5 μm (**a**) and 1 μm (**b**–**d**).

The exploration of the thermal stability of PTN-70 and PTN-71, conducted through thermogravimetric analysis (TGA), served as a pivotal dimension in the characterization of these materials. The TGA results, presented in the insightful Figure S5, unveiled a notable stability in both polymers, with PTN-70 exhibiting resilience up to temperatures of 400 °C and PTN-71 up to 350 °C. However, the intriguing observation during the TGA analysis surfaced at 150 °C, where a marginal decrease in mass was noted. This phenomenon was attributed to the persistence of solvent trapped within the micropore structure of the polymers. The solvent, proving recalcitrant to removal under vacuum conditions at room temperature, presented a significant challenge in achieving the desired purity of the materials. A subsequent treatment at 100 °C in a vacuum for 10 h proved to be an effective remedy, facilitating the complete removal of the persistent solvent within the micropore structure.

The meticulous examination of the pore properties of hypercrosslinked polymers (HCPs) unfolded with a thorough process involving desolvation at 120 °C for an extensive 10-h duration under vacuum, followed by nitrogen sorption and desorption analysis at a temperature of 77 K, as thoughtfully illustrated in Figure 3a. A critical aspect of this assessment was the scrutiny of the Brunauer–Emmett–Teller (BET) surface area, a pivotal parameter elucidating the accessibility of the internal surface of the polymer to nitrogen gas. The results portrayed a compelling narrative of contrast between PTN-70 and PTN-71, with the former exhibiting a notably larger surface area, measuring an impressive 1873 m^2 g^{-1}. In stark contrast, PTN-71 registered at 574 m^2 g^{-1}, indicating a considerable reduction in surface area compared to its counterpart. The Langmuir surface area, offering insights into monolayer coverage, accentuated this difference, with PTN-70 showcasing a Langmuir surface area of 2525 m^2 g^{-1}, surpassing the 770 m^2 g^{-1} observed for PTN-71 (Figures S6 and S7). Moreover, the exploration extended to the determination of the total pore volume, a pivotal metric influencing the storage capacity of gases within the material. The discernment that PTN-70 displayed a notably higher total pore volume, measuring 1.20 cm^3 g^{-1}, compared to the 0.40 cm^3 g^{-1} observed for PTN-71, indicated that PTN-70 harbors a more extensive network of internal voids, potentially offering enhanced storage capacity for gases.

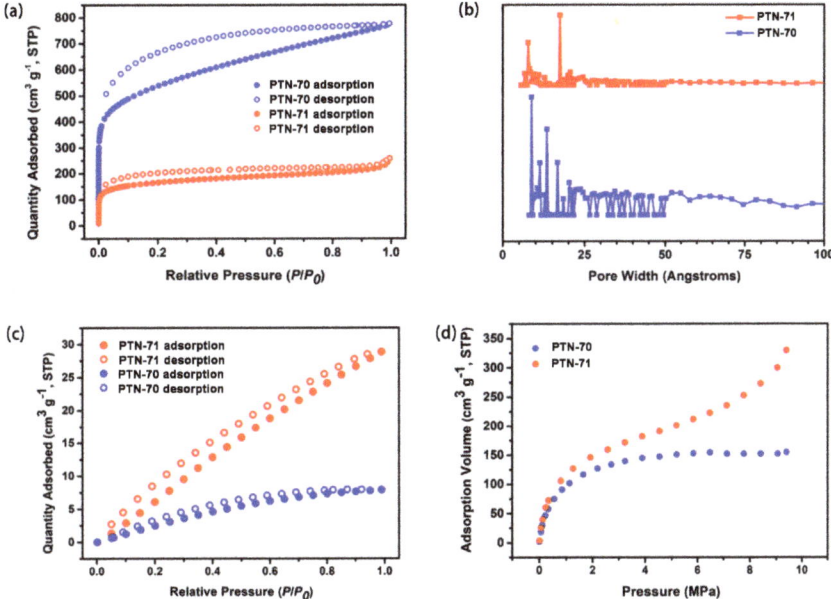

Figure 3. Nitrogen sorption isotherms of PTN-70 and PTN-71 at 77 K (**a**), pore size distributions calculated using NLDFT methods (**b**). (**c**) Methane (273 K) isotherm of PTN-70 and PTN-71. (**d**) Methane adsorption of PTN-70 and PTN-71 up to 95 bars at 273 K.

A pivotal consideration in this nuanced evaluation was the integration of a flexible linker within PTN-71, specifically 1,3-dichloropropane. This flexible linker has the potential to shape the creation of denser structures within the polymer, potentially influencing the diffusion of nitrogen gas within the material. This observation resonated harmoniously with the findings from morphological characterization, providing a plausible explanation for the reduction in both surface area and pore volume observed in PTN-71 compared to PTN-70. The nitrogen adsorption isotherms delved further into the intricacies of the pore structures of PTN-70 and PTN-71. A discernible surge in nitrogen gas uptake at low relative pressure ($P/P_0 < 0.001$) underscored the presence of abundant micropore structures within the HCPs. This phenomenon reflects the materials' adeptness at effectively adsorbing gas molecules within their intricate networks of sub-nanometer-sized pores. The observed hysteresis with higher capacity hinted at the irreversible uptake of gas molecules at mesopores or through pore entrances, highlighting the swelling behavior intrinsic to the polymer skeleton—a recurrent trait in amorphous microporous polymer networks. Such insights accentuate the dynamic nature of these materials and their capacity to undergo structural changes in response to gas adsorption.

The investigation was further enriched through nonlocal density functional theory (NLDET) utilizing the Micromeritics ASAP 2020 volumetric adsorption analyzer. This analytical approach served to corroborate the prevalence of micropores as the predominant structural component within the HCPs, as visually exemplified in Figure 3b. The analysis provided a comprehensive visualization of the pore size distributions of PTN-70 and PTN-71, predominantly falling within the microporous range, with dimensions consistently below 20 Å. This unequivocally solidified their classification as microporous polymers, boasting intricate networks of sub-nanometer-sized pores. However, a pivotal distinction surfaced between PTN-70 and PTN-71 in terms of mesoporous characteristics. PTN-70 exhibited the presence of mesopores within the range of 2–50 nm, contributing to a more diverse pore structure. In contrast, PTN-71 did not manifest such mesoporous features. This divergence can be attributed to the influential role played by the flexible linker, 1,3-dichloropropane. The flexible nature of this linker facilitated a more compact packing of polymer chains within PTN-71, as evidenced by the FE-SEM and TEM analyses, resulting in a discernible decrease in mesoporous features. This nuanced insight underscores the intricate interplay between the choice of linker and the resulting pore characteristics in hypercrosslinked polymers, shedding light on the material's ability to tailor its structural attributes based on the chemical composition.

In the assessment of natural gas (NG) adsorption performance for the two distinct polymers, PTN-70 and PTN-71, rigorous experiments were conducted by subjecting them individually to pure methane gas at a controlled temperature of 273 K. As shown in Figure 3c, despite the fact that the BET surface area of PTN-71 is obviously inferior to that of PTN-70, the methane adsorption capacity is much better, reaching 28.8 cm^3 g^{-1} at 273 K/bar, while the methane adsorption capacity of PTN-70 is 7.9 cm^3 g^{-1} at 273 K/bar. A pronounced hysteresis effect was observed in atmospheric methane adsorption and desorption measurements, corroborating the role of the flexible network structure in this distinct adsorption behavior. The methane adsorption isotherms unveiled intriguing and disparate behaviors. Upon increasing the pressure to 95 bars, PTN-70 exhibited a rapid and substantial increase in methane adsorption, swiftly ascending from 0 to 20 bars, whereafter the adsorption capacity reached a plateau, remaining relatively constant despite the continuous elevation of gas pressure (Figure 3d). In stark contrast, the methane adsorption isotherm for PTN-71 displayed a distinctive two-step pattern. After an initial phase, methane uptake experienced a sharp resurgence when the pressure reached 45 bars, intriguingly continuing to rise even at high pressures and ultimately saturating at 95 bars. This intriguing behavior was attributed to the swelling response of the flexible network structure within PTN-71 when subjected to elevated pressure conditions. Cycling tests were performed to assess the methane adsorption capabilities of two different polymers, PTN-70 and PTN-71, revealing their consistent and stable methane adsorption capacities (Figure S8). Importantly, this behavior provided a plausible explanation for why PTN-71, despite having a significantly lower surface area, exhibited a notably higher methane adsorption capacity under high gas pressure compared to PTN-70 [15,21].

An exploration into the properties of PTN-70 and PTN-71 has yielded a significant and intriguing revelation—their shared ability for reversible methane storage. This finding carries profound implications within the domain of adsorbed natural gas (ANG) technology, challenging longstanding assumptions in the field. Traditionally, the prevailing belief asserted a direct correlation between methane adsorption capacity and specific surface area, as well as total pore volume. Higher values of these parameters were generally expected to result in superior gas storage performance, as illustrated in Figure 4. However, PTN-71 has defied this conventional wisdom, emerging as a compelling exception to the established trend. Contrary to expectations, it exhibited remarkable reversible methane storage capabilities despite not conforming to the anticipated correlation between methane adsorption capacity, specific surface area, and total pore volume. This unexpected behavior prompts a reevaluation of existing paradigms in the field of ANG technology, challenging the notion that higher specific surface area and total pore volume invariably lead to enhanced gas storage performance [35–40].

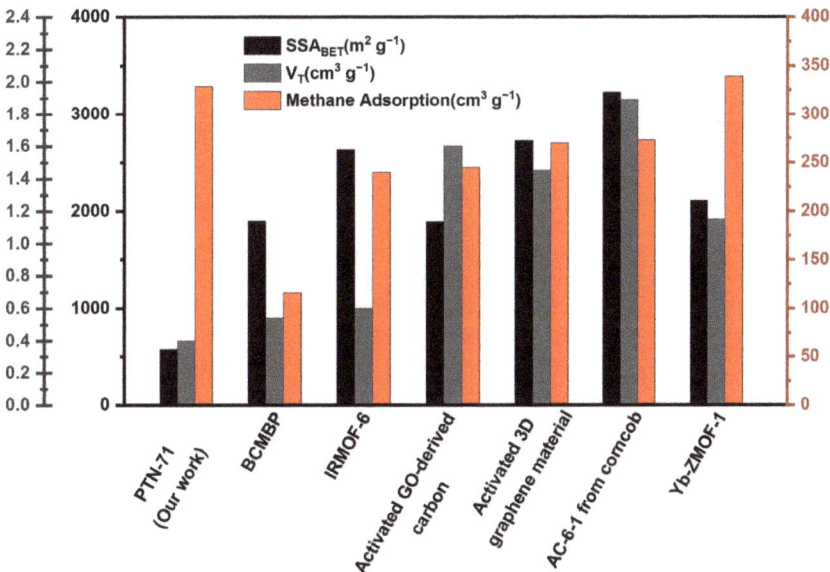

Figure 4. Adsorption capacity of methane for various materials recently reported. Details are located in Table S1 of the Supplementary Materials.

An illuminating point of comparison arises when juxtaposing the findings with the recent material Yb-ZMOF-1, celebrated for its exceptionally specific surface area, surpassing 1000 m^2 g^{-1} [40]. Intriguingly, PTN-71 has exhibited a methane adsorption capacity comparable to that of Yb-ZMOF-1, despite possessing a notably lower specific surface area and pore volume. This revelation serves as a striking testament to the influential role played by the flexible network structure within PTN-71. The surprising parity in methane adsorption between PTN-71 and Yb-ZMOF-1, despite substantial differences in structural characteristics, underscores the pivotal importance of the inherent flexibility within PTN-71. This flexibility appears to be a key factor contributing to its efficiency as a methane adsorbent, particularly under conditions of elevated gas pressure. The traditional paradigm linking superior gas storage performance to higher specific surface area and pore volume is thus challenged by the exceptional behavior exhibited by PTN-71.

The unprecedented methane storage capabilities demonstrated by PTN-71 challenge conventional assumptions within the field, offering a fresh perspective that accentuates the crucial role played by material flexibility and structure in methane adsorption. This insight opens novel avenues for the design of highly effective gas storage materials, suggesting that flexibility may be a crucial factor in optimizing performance under varying conditions. This groundbreaking discovery not only advances our understanding of methane adsorption mechanisms but also holds promise for the development of more efficient, flexible, and adaptable materials for adsorbed natural gas (ANG) technology [30].

4. Conclusions

In summary, our study harnessed the unique properties of triptycene as rigid building blocks to judiciously craft two distinct physicochemically stable triptycene-based hypercrosslinked polymers, denoted as PTN-70 and PTN-71, employing a cost-effective and expedient synthesis process under mild conditions. A comparative analysis between PTN-70, characterized by a rigid methylene linker, and PTN-71, featuring a flexible propylene linker, revealed intriguing findings. Despite PTN-71 exhibiting a lower BET surface area in contrast to the rigidly linked PTN-70, it demonstrated a remarkable and unexpected superiority in methane adsorption performance. The methane adsorption capacity of PTN-71 significantly outperformed that of PTN-70, reaching an impressive 329 $cm^3\ g^{-1}$, particularly under high-pressure conditions. This exceptional performance can be attributed to the unique flexible network structure of PTN-71, which undergoes a swelling response when subjected to elevated pressure conditions, facilitating enhanced methane adsorption. The flexibility introduced by the propylene linker played a pivotal role in expanding the porous network of PTN-71, presenting a stark contrast to the rigid network of PTN-70. This breakthrough challenges conventional assumptions by highlighting the potential of flexible-linked hypercrosslinked polymers (HCPs) as highly efficient methane adsorbents. The study not only expands the understanding of gas storage materials but also opens new avenues for advanced methane storage technologies with substantial implications for clean energy and environmental sustainability. The remarkable methane adsorption performance of PTN-71 underscores the versatility and adaptability of hypercrosslinked polymers, offering a promising avenue for the development of efficient and environmentally friendly methane storage solutions that can contribute significantly to the transition towards cleaner energy sources and sustainable practices.

Supplementary Materials: The following supporting information can be downloaded at: https://www.mdpi.com/article/10.3390/polym16010156/s1, Figure S1: FT-IR spectroscopy of triptycene, PTN-70 and PTN-71; Figure S2: ^{13}C MAS NMR spectra of (a) PTN-70 (blue) and (b) PTN-71 (red); Figure S3: SEM and TEM images of PTN-70 (a and b) and PTN-71 (c and d). Scale bar: 0.2 μm; Figure S4: Powder X-ray diffraction spectra of (a) PTN-70 and (b) PTN-71; Figure S5: Thermogravimetric analysis of PTN-70 and PTN-71 at room temperature (a) and after been treated at 100 °C in the vacuum for 10 h (b); Figure S6: The Brunauer-Emmett-Teller (BET) surface areas of PTN-70 calculated from nitrogen sorption analysis at 77 K; Figure S7: The Brunauer-Emmett-Teller (BET) surface areas of PTN-70 calculated from nitrogen sorption analysis at 77 K; Figure S8: The cycling test of methane adsorption of PTN-70 and PTN-71 up to 95 bars at 273 K; Table S1: Adsorption capacity of methane for various materials recently reported. References [35–40] are cited in the Supplementary Materials.

Author Contributions: Z.W.: Investigation, Methodology, Formal analysis, Validation, Writing, and Funding acquisition. F.G., H.M. and B.-B.Y.: Investigation, Methodology, Formal analysis, and Validation. X.-G.M. and J.-H.B.: Conceptualization, Formal analysis, Supervision. C.Z.: Conceptualization, Formal analysis, Supervision, Funding acquisition, and Writing and editing. F.G., H.M. and B.-B.Y. All authors have read and agreed to the published version of the manuscript.

Funding: This research was supported by the National Natural Science Foundation of China (22005110 and 22275062).

Institutional Review Board Statement: Not applicable.

Data Availability Statement: Research data are available in this document and in the related Supplementary Materials submitted.

Acknowledgments: We thank the Analytical and Testing Center of Wuhan Textile University for the related analysis.

Conflicts of Interest: The authors declare no conflict of interest.

References

1. Menon, V.C.; Komarneni, S. Porous Adsorbents for Vehicular Natural Gas Storage: A Review. *J. Porous Mater.* **1998**, *5*, 43–58. [CrossRef]
2. Zou, X.Y.; Xue, R.; An, Z.W.; Li, H.W.; Zhang, J.L.; Jiang, Y.; Huang, L.J.; Wu, W.; Wang, S.F.; Hu, G.-H.; et al. Recent Advances in Flexible CNC-Based Chiral Nematic Film Materials. *Small* **2023**, *2023*, 2303778. [CrossRef] [PubMed]
3. Xue, R.; Zhao, H.; An, Z.W.; Wu, W.; Jiang, Y.; Li, P.; Huang, C.-X.; Shi, D.; Li, R.K.Y.; Hu, G.-H.; et al. Self-healable, solvent response cellulose nanocrystal/ waterborne polyurethane nanocomposites with encryption capability. *ACS Nano* **2023**, *17*, 5653–5662. [CrossRef]
4. Connolly, B.M.; Madden, D.; Wheatley, A.; Fairen-Jimenez, D. Shaping the Future of Fuel: Monolithic Metal-Organic Frameworks for High-Density Gas Storage. *J. Am. Chem. Soc.* **2020**, *142*, 8541–8549. [CrossRef] [PubMed]
5. Rowland, C.A.; Lorzing, G.R.; Gosselin, E.J.; Trump, B.A.; Yap, G.P.A.; Brown, C.M.; Bloch, E.D. Methane Storage in Paddlewheel-Based Porous Coordination Cages. *J. Am. Chem. Soc.* **2018**, *140*, 11153–11157. [CrossRef] [PubMed]
6. Acharyya, K.; Mukherjee, P.S. Organic Imine Cages: Molecular Marriage and Applications. *Angew. Chem. Int. Ed.* **2019**, *58*, 8640–8653. [CrossRef]
7. Bracco, S.; Piga, D.; Bassanetti, I.; Perego, J.; Comotti, A.; Sozzani, P. Porous 3D polymers for high pressure methane storage and carbon dioxide capture. *J. Mater. Chem. A* **2017**, *5*, 10328–10337. [CrossRef]
8. Che, S.; Pang, J.D.; Kalin, A.J.; Wang, C.X.; Ji, X.Z.; Li, J.L.; Cole, D.; Li, J.L.; Tu, X.M.; Zhang, Q.; et al. Rigid Ladder-Type Porous Polymer Networks for Entropically Favorable Gas Adsorption. *ACS Mater. Lett.* **2020**, *2*, 49–54. [CrossRef]
9. Zhang, A.J.; Zhang, Q.K.; Bai, H.; Li, L.; Li, J. Polymeric nanoporous materials fabricated with supercritical CO_2 and CO_2-expanded liquids. *Chem. Soc. Rev.* **2014**, *43*, 6938–6953. [CrossRef]
10. Zhu, T.T.; Pei, B.Y.; Di, T.; Xia, Y.X.; Li, T.S.; Li, L. Thirty-minute preparation of microporous polyimides with large surface areas for ammonia adsorption. *Green Chem.* **2020**, *22*, 7003–7009. [CrossRef]
11. Makal, T.A.; Li, J.R.; Lu, W.G.; Zhou, H.C. Methane storage in advanced porous materials. *Chem. Soc. Rev.* **2012**, *41*, 7761–7779. [CrossRef] [PubMed]
12. Liu, Y.Z.; Li, S.; Dai, L.; Li, J.N.; Lv, J.N.; Zhu, Z.J.J.; Yin, A.X.; Li, P.F.; Wang, B. The Synthesis of Hexaazatrinaphthylene-Based 2D Conjugated Copper Metal-Organic Framework for Highly Selective and Stable Electroreduction of CO_2 to Methane. *Angew. Chem. Int. Ed.* **2021**, *60*, 16409–16415. [CrossRef] [PubMed]
13. Kumar, K.V.; Preuss, K.; Titirici, M.M.; Rodriguez-Reinoso, F. Nanoporous Materials for the Onboard Storage of Natural Gas. *Chem. Rev.* **2017**, *117*, 1796–1825. [CrossRef] [PubMed]
14. Fang, H.; Zheng, B.; Zhang, Z.H.; Li, H.X.; Xue, D.X.; Bai, J.F. Ligand-Conformer-Induced Formation of Zirconium-Organic Framework for Methane Storage and MTO Product Separation. *Angew. Chem. Int. Ed.* **2021**, *60*, 16521–16528. [CrossRef] [PubMed]
15. Zhang, L.; Sun, T.; Dong, Y.B.; Fang, Z.B.; Xu, Y.X. Electron-donating group induced rapid synthesis of hyper-crosslinked polymers. *Sci. Bull.* **2022**, *67*, 1416–1420. [CrossRef] [PubMed]
16. Tan, L.X.; Tan, B. Hypercrosslinked porous polymer materials: Design, synthesis, and applications. *Chem. Soc. Rev.* **2017**, *46*, 3322–3356. [CrossRef] [PubMed]
17. Lee, J.S.M.; Briggs, M.E.; Hu, C.C.; Cooper, A.I. Controlling electric double-layer capacitance and pseudocapacitance in heteroatom-doped carbons derived from hypercrosslinked microporous polymers. *Nano Energy* **2018**, *46*, 277–289. [CrossRef]
18. Liu, N.; Ma, H.; Sun, R.; Zhang, Q.-P.; Tan, B.; Zhang, C. Porous Triptycene Network Based on Tröger's Base for CO_2 Capture and Iodine Enrichment. *ACS Appl. Mater. Interfaces* **2023**, *15*, 30402–30408. [CrossRef]
19. Eckstein, B.J.; Brown, L.C.; Noll, B.C.; Moghadasnia, M.P.; Balaich, G.J.; McGuirk, C.M. A Porous Chalcogen-Bonded Organic Framework. *J. Am. Chem. Soc.* **2021**, *143*, 20207–20215. [CrossRef]
20. Soto, C.; Torres-Cuevas, E.S.; Gonzalez-Ortega, A.; Palacio, L.; Pradanos, P.; Freeman, B.D.; Lozano, A.E.; Hernandez, A. Hydrogen Recovery by Mixed Matrix Membranes Made from 6FCl-APAF HPA with Different Contents of a Porous Polymer Network and Their Thermal Rearrangement. *Polymers* **2021**, *13*, 4343. [CrossRef]
21. Zotkin, M.A.; Alentiev, D.A.; Shorunov, S.V.; Sokolov, S.E.; Gavrilova, N.N.; Bermeshev, M.V. Microporous polynorbornenes bearing carbocyclic substituents: Structure-property study. *Polymer* **2023**, *269*, 125732. [CrossRef]
22. Mason, J.A.; Oktawiec, J.; Taylor, M.K.; Hudson, M.R.; Rodriguez, J.; Bachman, J.E.; Gonzalez, M.I.; Cervellino, A.; Guagliardi, A.; Brown, C.M.; et al. Methane storage in flexible metal-organic frameworks with intrinsic thermal management. *Nature* **2015**, *527*, 357–361. [CrossRef] [PubMed]
23. Rozyyev, V.; Thirion, D.; Ullah, R.; Lee, J.; Jung, M.; Oh, H.; Atilhan, M.; Yavuz, C.T. High-capacity methane storage in flexible alkane-linked porous aromatic network polymers. *Nat. Energy* **2019**, *4*, 604–611. [CrossRef]
24. Llewellyn, P.L.; Bourrelly, S.; Serre, C.; Filinchuk, Y.; Férey, G. How Hydration Drastically Improves Adsorption Selectivity for CO_2 over CH_4 in the Flexible Chromium Terephthalate MIL-53. *Angew. Chem. Int. Ed.* **2006**, *45*, 7751–7754. [CrossRef] [PubMed]
25. Horike, S.; Shimomura, S.; Kitagawa, S. Soft porous crystals. *Nat. Chem.* **2009**, *1*, 695–704. [CrossRef]

26. Zhan, Z.; Yu, J.C.; Li, S.Q.; Yi, X.X.; Wang, J.Y.; Wang, S.L.; Tan, B. Ultrathin Hollow Co/N/C Spheres from Hyper-Crosslinked Polymers by a New Universal Strategy with Boosted ORR Efficiency. *Small* **2023**, *19*, 2207646. [CrossRef]
27. Qiao, S.; Li, Z.; Zhang, B.; Li, Q.; Jin, W.; Zhang, Y.; Wang, W.; Li, Q.; Liu, X. Flexible chain & rigid skeleton complementation polycarbazole microporous system for gas storage. *Micropor. Mesopor. Mater.* **2019**, *284*, 205–211.
28. An, Z.-W.; Ye, K.; Xue, R.; Zhao, H.; Liu, Y.; Li, P.; Chen, Z.-M.; Huang, C.-X.; Hu, G.-H. Recent advances in self-healing polyurethane based on dynamic covalent bonds combined with other self-healing methods. *Nanosacle* **2023**, *15*, 7591. [CrossRef]
29. Li, H.-W.; Zhang, J.-L.; Xue, R.; An, Z.-W.; Wu, W.; Liu, Y.; Hu, G.-H.; Zhao, H. Construction of self-healable and recyclable waterborne polyurethane-MOF membranes for adsorption of dye wastewater. *Sep. Purif. Technol.* **2023**, *320*, 12415. [CrossRef]
30. Dai, L.; Dong, A.W.; Meng, X.J.; Liu, H.Y.; Li, Y.T.; Li, P.F.; Wang, B. Enhancement of Visible-Light-Driven Hydrogen Evolution Activity of 2D pi-Conjugated Bipyridine-Based Covalent Organic Frameworks via Post-Protonation. *Angew. Chem. Int. Ed.* **2023**, *475*, 146264. [CrossRef]
31. Li, M.P.; Ren, H.; Sun, F.X.; Tian, Y.Y.; Zhu, Y.L.; Li, J.L.; Mu, X.; Xu, J.; Deng, F.; Zhu, G.S. Construction of Porous Aromatic Frameworks with Exceptional Porosity via Building Unit Engineering. *Adv. Mater.* **2018**, *30*, 1804169. [CrossRef] [PubMed]
32. Zu, Y.C.; Li, J.W.; Li, X.L.; Zhao, T.Y.; Ren, H.; Sun, F.X. Imine-linked porous aromatic frameworks based on spirobifluorene building blocks for CO_2 separation. *Micropor. Mesopor. Mater.* **2022**, *334*, 111779. [CrossRef]
33. Zhang, Q.P.; Wang, Z.; Zhang, Z.W.; Zhai, T.L.; Chen, J.J.; Ma, H.; Tan, B.; Zhang, C. Triptycene-based Chiral Porous Polyimides for Enantioselective Membrane Separation. *Angew. Chem. Int. Ed.* **2021**, *60*, 12781–12785. [CrossRef] [PubMed]
34. Wang, Z.; Ma, H.; Zhai, T.L.; Cheng, G.; Xu, Q.; Liu, J.M.; Yang, J.K.; Zhang, Q.M.; Zhang, Q.P.; Zheng, Y.S.; et al. Networked Cages for Enhanced CO_2 Capture and Sensing. *Adv. Sci.* **2018**, *5*, 1800141. [CrossRef]
35. Wood, C.D.; Tan, B.; Trewin, A.; Su, F.; Rosseinsky, M.J.; Bradshaw, D.; Sun, Y.; Zhou, L.; Cooper, A.I. Microporous Organic Polymers for Methane Storage. *Adv. Mater.* **2008**, *20*, 1916–1921. [CrossRef]
36. Eddaoudi, M.; Kim, J.; Rosi, N.; Vodak, D.; Wachter, J.; O'Keeffe, M.; Yaghi, O.M. Systematic Design of Pore Size and Functionality in Isoreticular MOFs and Their Application in Methane Storage. *Science* **2002**, *295*, 469–472. [CrossRef] [PubMed]
37. Srinivas, G.; Burress, J.; Yildirim, T. Graphene oxide derived carbons (GODCs): Synthesis and gas adsorption properties. *Energy Environ. Sci.* **2012**, *5*, 6453–6459. [CrossRef]
38. Mahmoudian, L.; Rashidi, A.; Dehghani, H.; Rahighi, R. Single-step scalable synthesis of three-dimensional highly porous graphene with favorable methane adsorption. *Chem. Eng. J.* **2016**, *304*, 784–792. [CrossRef]
39. Liu, B.S.; Wang, W.S.; Wang, N.; Au, C.T. Preparation of activated carbon with high surface area for high-capacity methane storage. *J. Energy Chem.* **2014**, *23*, 662–668. [CrossRef]
40. Li, H.X.; Zhang, Z.H.; Fang, H.; Xue, D.X.; Bai, J.F. Synthesis, structure and high methane storage of pure D6R Yb(Y) nonanuclear cluster-based zeolite-like metal-organic frameworks. *J. Mater. Chem. A* **2022**, *10*, 14795–14798. [CrossRef]

Disclaimer/Publisher's Note: The statements, opinions and data contained in all publications are solely those of the individual author(s) and contributor(s) and not of MDPI and/or the editor(s). MDPI and/or the editor(s) disclaim responsibility for any injury to people or property resulting from any ideas, methods, instructions or products referred to in the content.

Article

Recommended Values for the Hydrophobicity and Mechanical Properties of Coating Materials Usable for Preparing Controlled-Release Fertilizers

Yajing Wang [†], Juan Li [†], Ru Lin, Dianrun Gu, Yuanfang Zhou, Han Li and Xiangdong Yang *

State Key Laboratory of Efficient Utilization of Arid and Semi-arid Arable Land in Northern China/Key Laboratory of Plant Nutrition and Fertilizer, Ministry of Agriculture and Rural Affairs/the Institute of Agricultural Resources and Regional Planning, Chinese Academy of Agricultural Sciences, Beijing 100081, China; 18837129765@163.com (Y.W.); lijuan02@caas.cn (J.L.); mumuru666@163.com (R.L.); dianrunna@163.com (D.G.); zyf20220129@163.com (Y.Z.); qhxna2288@163.com (H.L.)
* Correspondence: yangxiangdong@caas.cn; Tel.: +86-10-82109614
[†] These authors contributed equally to this work.

Abstract: The hydrophobicity and mechanical properties of coating materials and the nitrogen (N) release rates of 11 kinds of controlled-release fertilizers (CRFs) were determined in this study. The results show that the N release periods of the CRFs had negative correlations with the water absorption (WA) of the coating materials ($y = 166.06x^{-1.24}$, $r = 0.986$), while they were positively correlated with the water contact angle (WCA) and elongation at break (EB) ($y = 37.28x^{0.18}$, $r = 0.701$; $y = -19.42 + 2.57x$, $r = 0.737$). According to the fitted functional equation, CRFs that could fulfil the N release period of 30 days had a coating material WA < 2.4%, WCA > 68.8°, and EB > 57.7%. The recommended values for a CRF that can fulfil the N release period of 30 days are WA < 3.0%, WCA > 60.0°, and EB > 30.0% in the coating materials. CRFs with different nutrient release periods can be designed according to the recommended values to meet the needs of different crops. Furthermore, our experiments have illustrated that the N release period target of 30 days can be reached for modified sulfur-coated fertilizers (MSCFs) by improving their mechanical properties.

Keywords: controlled-release fertilizer; coating material; hydrophobicity; mechanical property

Citation: Wang, Y.; Li, J.; Lin, R.; Gu, D.; Zhou, Y.; Li, H.; Yang, X. Recommended Values for the Hydrophobicity and Mechanical Properties of Coating Materials Usable for Preparing Controlled-Release Fertilizers. *Polymers* **2023**, *15*, 4687. https://doi.org/10.3390/polym15244687

Academic Editor: Ian Wyman

Received: 26 October 2023
Revised: 20 November 2023
Accepted: 20 November 2023
Published: 12 December 2023

Copyright: © 2023 by the authors. Licensee MDPI, Basel, Switzerland. This article is an open access article distributed under the terms and conditions of the Creative Commons Attribution (CC BY) license (https://creativecommons.org/licenses/by/4.0/).

1. Introduction

Food security strongly depends on a sufficient nutrient supply [1,2]. However, the excess use of commercial fertilizers for maximum agricultural crop production leads to serious resource consumption and severe environmental pollution [3,4], as well as directly or indirectly affecting human health [5]. A one-off application of a controlled-release fertilizer (CRF) can meet the nutrient requirements of crops [6]. It has been illustrated that the use of CRF is a green technology that not only improves fertilizer use efficiency, reduces nitrogen (N) loss, and saves labor [7], but also unifies these factors with its environmental benefits and economic benefits [8]. The development of CRFs mainly depends on the development process of coating materials, ranging from inorganic to organic and from natural polymers to organic synthetic polymers [9]. To date, only a few polymers, such as alkyd resin [10,11], polyethylene (PE) [12], polyurethane (PU) [13], and styrene–acrylic latex (SAL) [14], have exhibited excellent controlled-release effects. Varieties of inorganic, organic, and biological materials, such as sulfur (S), calcium magnesium phosphate rock powder, polyvinyl alcohol (PVA), natural cellulose, lignin, chitin, and tung oil, have various drawbacks [15]. For example, the film structure collapses when in contact with water, swelling or rupturing when absorbing the water due to its poor ability to hinder water and fertilizer; in turn, it exhibits poor controlled-release characteristics.

The nutrient release of CRFs has been found to be closely associated with their film structure [16–18] and the characteristics of the coating materials [6]. Previous studies have assumed that the film structure remains stable when a CRF is immersed in water under ideal conditions, and several models have been developed to explain the mechanism of nutrient diffusion based on this assumption [16,19,20]. However, establishing a comprehensive release mechanism is challenging due to uncertain external environmental factors and material properties. In reality, most film properties undergo changes upon contact with water [21], resulting in alterations in both film structure and nutrient release rate. Therefore, further improvements are needed for existing release models [22–24], particularly considering the complex release process of swelling film materials [25]. Consequently, it can be concluded that the performance of controlled-release coatings primarily depends on their material properties.

In the past, it was widely believed that the hydrophobicity [26,27] and mechanical properties [28] of coating materials were crucial factors in controlling nutrient release. For instance, the water absorption (WA) of a starch/PVA-CRF coating film exceeded 360% [29], while PVA/biochar composite coating materials exhibited WA values ranging from 200 to 300% [30]. Similarly, biodegradable starch/PVA/bentonite graft polymers displayed WA values within the range of 200–300% [31]. Furthermore, double-coated slow-release fertilizers developed with ethyl cellulose ether and starch-based highly absorbent polymers demonstrated approximately 100% WA [9]. These films had high hygroscopicity and rapidly absorbed water, resulting in reduced controllability. Therefore, efforts were made to enhance their controllability by reducing their WA. For example, a composite film modified with nano-silica and γ-polyglutamic acid achieved substantial reductions in its WA, along with enhanced densification compared to unmodified PVA film; this modification led to improved controllability [32]. Additionally, etherified epoxy resin was employed in modifying the PVA film, which resulted in lower WA (884 g/kg) than unmodified PVA at the same concentration; furthermore, a CRF coated with modified PVA exhibited a slow nutrient release performance [33].

The water contact angle (WCA) serves as a crucial indicator of the hydrophobic performance of materials, with larger WCA values indicating stronger water barrier properties [34]. However, bio-based materials exhibit smaller WCAs and shorter nutrient release periods. To optimize their water resistance, researchers have enhanced the characteristics of their coating materials. By modifying the hydrophobic properties of these coatings, it is possible to significantly increase the WCA and greatly improve the controlled-release performance. For instance, Xie et al. developed a biomimetic bio-based CRF by incorporating a superhydrophobic film surface modified with micro/nanoscale diatomaceous earth silica. This modification reduced the nutrient release rate and extended the nutrient release period twofold compared to an unmodified CRF [35]. The surface energy of the bio-based CRF was reduced and the WCA increased from 89.8 to 158.9° after modification, resulting in an extended nutrient release period of the modified CRF with a 3% coating rate from 5 days to approximately 28 days [36]. Additionally, the hydrophobicity of PU was enhanced through the use of paraffin wax as well as nano-silica fillers or by incorporating hydroxypropyl-capped polydimethylsiloxane (HP-PDMS) into the PU reaction [37–39]. Furthermore, when nano-cellulose crystals were added to PVA with high WA, the WCA of PVA increased from 29.3 to 69.7° and led to an approximately twofold extension in the nutrient release period for the modified CRF [40]. These findings demonstrate that hydrophobic modifications in coating materials can significantly enhance the nutrient release performance of CRFs.

Usually, materials with simultaneous good hydrophobicity and poor mechanical properties often lack controllability. In most coating processes, nanomaterials or other modifiers are employed to enhance the mechanical properties of coatings. For instance, when S was utilized as a coating to derive sulfur-coated fertilizers (SCFs) with a WCA of 78°, it exhibited favorable water resistance; however, the SCFs had inadequate mechanical properties. By modifying S with dicyclopentadiene (DCPD), the compressive strength increased from 27 to 47 N, and the 7-day cumulative release rate of N from the SCFs decreased from 83 to 54% [41]. Consequently, the mechanical modification of S to yield good hydrophobic properties resulted in an improvement in its controlled-release performance. Furthermore, for polymeric materials, enhancing their mechanical properties can also lead to improved controlled-release characteristics. For example, silica-modified PU demonstrated an increase in tensile strength (TS) from 40 to 54 MPa while decreasing its elongation at break (EB) from 7 to 2%, thereby extending the nutrient release period by approximately 32% [42]. Moreover, compared to the pure PU material, incorporating modified natural pyrophyllite into a PU composite enhanced both its TS and EB by approximately 0.61 MPa and 42%, respectively, and increased the nutrient release period of the CRF by an additional 15 days [43].

The unmodified materials mentioned above all exhibited drawbacks in terms of hydrophobicity and mechanical properties, resulting in poor controllability and the need for improvements in their nutrient release periods through targeted modifications. However, it was found that unmodified PE with enhanced hydrophobicity and mechanical properties could achieve an N release period of eight months [17]. Therefore, achieving the good controllability of coating materials requires ensuring, at least to some degree, their hydrophobic and mechanical characteristics simultaneously. In other words, there exists an intrinsic relationship between the hydrophobicity and mechanical properties of the coating materials and the nutrient release period of the CRF. However, it remains unclear to what extent the hydrophobicity and mechanical properties of the coating materials need to be optimized to ensure good controllability. Thus, we hypothesized that (1) a correlation can be established between the hydrophobicity and mechanical properties of coating materials and the nutrient release period of CRFs; (2) recommended values for an excellent controlled-release performance can be determined for both hydrophobicity and the mechanical properties; (3) these recommended values can be experimentally validated for reliability. The objective of this study was to determine the values of both the hydrophobicity and mechanical properties that are necessary for preparing CRFs using suitable coating materials. Firstly, we measured the films' hydrophobicity, mechanical properties, and N release periods based on existing CRF products. Subsequently, we established a numerical fitting method to establish relationships between these parameters. Secondly, we calculated the parameter values corresponding to N release periods of 30 or 90 days, respectively. Finally, the reliability of the relational model was validated through enhancements in the physical properties of the materials, ensuring their adherence to recommended values. This relational model can effectively determine the mechanical properties and hydrophobic value required for achieving an exceptional release performance in CRFs, thereby offering valuable parameter guidance for coating material development.

2. Materials and Methods

2.1. Experimental Materials, Equipment, and Devices

Urea granules (2–4 mm diameter, 46% nitrogen content, Shandong Hualu-Hengsheng Group Co., Ltd., Dezhou, China); polypropylene glycol (PPG, molecular weight (MW) = 400, Sinopec Asset Management Co., Ltd., Tianjin Petrochemical Branch, Tianjin, China); 4,4'-diphenylmethane diisocyanate (MDI, Wanhua Chemical Group Co., Ltd., Yantai, China); polytetrahydrofuran (PTMG, MW = 250, Xuzhou Yihui Yang New Material Co., Ltd., Xuzhou, China); polyethylene glycol (PEG, MW = 200, Lotte Chemical, Seoul, Republic of Korea); polycapro-lactone (PCL1,MW = 500; PCL2, MW = 1000, Hunan Juren Chemical New Material Technology Co., Ltd., Changsha, China); poly(1,4'-butylene adipate) (PBA, MW = 1000, Shandong Jiaying Chemical Co., Ltd., Qingdao, China); polyvinyl alcohol (PVA, MW = 1750, Shanghai Maclean Biochemical Technology Co., Ltd., Shanghai, China); polyethylene (PE, Sinopec Beijing Yanshan Petrochemical Co., Beijing, China); styrene-acrylic latex 1 (SAL1, PRIMAL AS-2010, Dow Chemical Co., Shanghai, China); styrene-acrylic latex 2 (SAL2, PRIMAL AS-8098, Dow Chemical Co., Shanghai, China); sulfur (S, Beijing Jinyuanteng Trading Co., Ltd., Beijing, China); sulfur-coated fertilizer (SCF, Shanghai Hanfeng Slow Release Fertilizer Co., Ltd., Shanghai, China); 4-dimethylaminobenzaldehyde (PDAB, AR, Shanghai Maclean Biochemical Technology Co., Ltd., Beijing, China); nano-silica (SiO_2, Jiangsu Xianfeng Nanomaterials Technology Co., Ltd., Nanjing, China).

Microcomputer heating platform (GVECTECH V3030T, Dingxinyi Experimental Equipment Co., Ltd., Shanghai, China); air-drying oven (Shanghai Lichen Instrument Technology Co., Ltd., Shanghai, China); ultraviolet spectrophotometer (SPECORD200, Analytik Jena, Jena, Germany); electronic balance (ME3002, Mettler Toled, Germany); spiral micrometer (0–25 mm, 0.001 mm, Nanjing Sutech Measuring Instruments Co., Ltd., Nanjing, China); contact angle measuring instrument (SCA20, Data Physics, Filderstadt, Germany); vertical computer servo material testing machine (CREE-8003A, Dongguan Krui Instrument Co., Ltd., Guangzhou, China); small fluidized bed (New Fertilizer Experimental Base of International Agricultural High-tech Industrial Park, Chinese Academy of Agricultural Sciences, self-made, Bejing, China); coating pan (YB400, Henan Qineng Machinery and Equipment Co., Ltd., Zhengzhou, China).

2.2. Experimental Methods

2.2.1. Preparation of Films

Preparation of PVA film: A PVA solution was prepared by completely dissolving 2.95 g of PVA granules in 46 mL of deionized water at 100 °C, followed by the addition of 1.05 g of polyvinyl pyrrolidone under stirring at 90 °C for 2 h. Subsequently, the resulting PVA coating solution was poured onto a Teflon board (dimensions: 25 cm × 25 cm × 0.1 mm) and allowed to flow naturally. The films were then placed on a microcomputer heating platform at 50 °C for a reaction time of 2 h. Finally, the PVA films were obtained after drying them in an air-drying oven at 50 °C for a duration of 12 h (Figure 1).

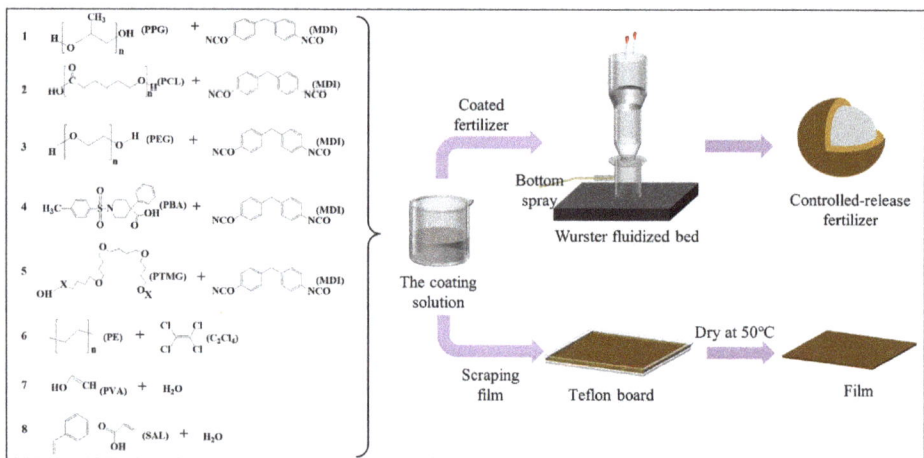

Figure 1. Schematic diagram of the preparation process of CRF and film. Note: NO. 1–5 are the formulations of five PU coating solutions (PPG, PCL, PEG, PBA and PTMG); NO. 6 is the formulations of PE coating solutions; NO. 7 is the formulations of PVA coating solutions; NO. 8 is the formulations of SAL coating solutions.

Preparation of SAL1 film: A 30% dry matter content SAL1 solution was prepared by adding 66.7 mL of water to 100 g of SAL1 with a 50% dry matter content. After thorough mixing, the mixture was poured onto a Teflon board (25 cm × 25 cm × 0.1 cm) and allowed to flow naturally before being placed on a micro-computer heating platform at 50 °C for 2 hours. The SAL1 film was obtained after drying the films in an air-drying oven at 50 °C for 12 hours (Figure 1).

Preparation of PE film: A PE solution with a concentration of 8% (w/w) was prepared by dissolving PE granules into a heated tetrachloroethylene solution at 80–90 °C, resulting in a homogeneous mixture. The mixture was then stirred and raised to 120 °C, maintaining this temperature for 20–30 min to ensure complete dissolution of the PE granules. Subsequently, the resulting PE coating solution was poured onto a Teflon board, allowing for the formation of a PE film through the complete volatilization of tetrachloroethylene.

Preparation of PU films: The PU films were prepared using a one-step synthetic method. Polyol (PPG, PTMG, PBA, PEG, and PCL) and MDI were weighed according to a molar ratio of 1:1 (Table 1) and thoroughly mixed. The mixture was then poured onto a Teflon board (dimensions: length × width × depth = 25 cm × 25 cm × 0.1 cm) to allow natural flow. Subsequently, the films were placed on a micro-computer heating platform at 50 °C for 2 h to initiate the reaction. Following drying at 50 °C for an additional 12 h in an air-drying oven, six types of PU films (PPG, PTMG, PBA, PEG, PCL1, and PCL2) were obtained based on the formula for PU coating materials.

Table 1. Formula scheme of CRFs.

NO.	Treatment	Quantity of Coating Materials	Urea (kg)	Coating Rate (%)	Temperature (°C)
1	PVACF	1000 g PVA solution (5% w/w)	1	5	80
2	SALCF1	267 g SAL1 solution (30% w/w)	1	8	45
3	SALCF2	267 g SAL2 solution (30% w/w)	1	8	45
4	PECF	625 g PE solution (8% w/w)	1	5	85
5	PPGCF	19.96 g PPG, 20.04 g MDI	1	4	70
6	PTMGCF	19.88 g PTMG250, 20.21 MDI	1	4	70
7	PBACF	30.8 g PBA 1000, 9.2 g MDI	1	4	70
8	PEGCF	18 g PEG200, 23.8 g MDI	1	4	70
9	PCLCF1	25.76 g PCL3050, 16.6 g MDI	1	4	70
10	PCLCF2	32 g PCL2105, 8 g MDI	1	4	70

Note: The coating rate is the weight ratio of the coating materials (without solvent) to the urea.

2.2.2. Preparation of CRFs in the Fluidized Bed Spray-Coating Process

The Wurster fluidized bed was employed as the coating apparatus (Figure 1), and diverse materials were utilized to configure the coating solution in accordance with the methodology outlined in Section 2.2.1, while a comprehensive scheme detailing material proportions can be found in Table 1.

Preparation of PVA-coated CRF (PVACF): The process flow chart for preparing PVACF is presented in Figure 1. Initially, the temperature of the coated fluidized bed was preheated to 80 °C, and large urea granules were added based on the formula design provided in Table 1 until a steady state of fluidization was achieved. Subsequently, liquid paraffin was introduced into the fluidized bed and mixed with the urea particles for 2 min. PVACF was then prepared by adding a PVA solution with a concentration of 5%, which was controlled using a peristaltic pump and pumped into the two-fluid nozzle at the bottom of the fluidized bed at a rate of 30 mL/min. The solution was atomized and sprayed onto the surface of the urea particles to coat them, resulting in the CRF's preparation.

Preparation of SAL-coated CRF (SALCF): The process flow chart for the preparation of SALCF is illustrated in Figure 1. The materials were added based on the formula design provided in Table 1, and both SALCF1 and SALCF2 were prepared following the same procedure as PVACF.

Preparation of PE-coated CRF (PECF): The process flow chart of preparing the PECF is shown in Figure 1. The materials were added according to the formula design in Table 1 and the PECF was prepared in the same way as PVACF [44].

Preparation of PU-coated CRFs (PUCFs): The PUCFs were prepared using the in situ reaction method [37]. The process flow chart for preparing the PUCFs is illustrated in Figure 1. Initially, the coated fluidized bed was preheated to 70 °C, and large urea granules were added according to the formula design specified in Table 1 until a steady state of fluidization was achieved. Subsequently, liquid paraffin was introduced into the fluidized bed and mixed with the urea granules for a duration of 2 min. The PUCFs were then synthesized through a dip-coating process involving polyol (PPG, PTMG, PBA, PEG, PCL1, and PCL2) and MDI addition. The reaction proceeded within the fluidized bed for approximately 20 min before removing the resulting PUCFs. This procedure yielded six distinct types of PUCFs.

2.2.3. Preparation of Modified S Film and Modified SCFs (MSCF)

Preparation of modified S film: S powder and PCL1 were accurately weighed to the weight ratios of MSCF1—S:PCL1:MDI = 15:9.91:5.09 and MSCF2—S:PCL1:MDI = 15:4.96:2.54 and thoroughly mixed to achieve homogeneity. Subsequently, MDI was added to the mixture to formulate the coating materials, which were then applied onto a flat film using a steel wire rod coater. A round pot coating machine containing 500 g of large-particle urea was preheated at 75 °C for 2 min. The atomizing nozzle was employed to spray the blend of PCL1 and S powder onto the surface of the large-particle urea while simultaneously introducing MDI into the coating pan for the urea coating operation.

Preparation of MSCF: The round pot coating machine was preheated at 75 °C for 2 min with the addition of 500 g of large-particle urea. A mixture of S powder and PCL1 was sprayed onto the surface of the large-particle urea using an atomizing nozzle, while MDI was introduced into the coating pan for the urea coating operation.

2.2.4. Determination of the N Release Period of the CRFs

The N release period of the CRFs was determined using the water immersion method at 25 °C. The content of urea–N in water was spectrophotometrically determined in accordance with regulation ISO-18644-2016 [45]. The nutrient release index formula is as follows:

$$\eta_{t1} = \frac{m_{t1}}{M} \times 100\% \quad (1)$$

$$\eta = \frac{\sum \eta_m}{M} \times 100\% \quad (2)$$

$$\eta_{\Delta t} = \frac{\eta_{tn} - \eta_{t1}}{t_n - t_1} \times 100\% \quad (3)$$

$$T = 1 + \frac{80\% - \eta_{t1}}{\eta_{\Delta t}} \quad (4)$$

η_t is the N cumulative release rate; t_1 is the first day; t_n is n days after incubation; η_{t1} is the initial N release rate; η_{tn} is the N cumulative release rate on the nth day; m_{t1} is the N content released on the 1st day; η_m is the N content released over n days; M is the total N content released in the CRFs; $\eta_{\Delta t}$ is the average N release rate; and T is the N release period.

2.2.5. Measurement of the WA of Films

The WA of films could be evaluated using the gravimetric method according to GB/T 1034-2008. Dried samples were weighed in triplicate, then immersed in 200 mL distilled water at 23 ± 1 °C for 24 h. The samples, taken out using tweezers, were weighed after filtering, and the WA of the films was calculated based on Equation (5):

$$WA = \frac{M_2 - M_1}{M_1} \times 100\% \quad (5)$$

where M_1 and M_2 refer to the dry and wet weights of the films, respectively.

2.2.6. Measurement of the WCA of Films

The contact angle, ranging from 0 to 180°, represents the angle formed at the interface of the solid, gas, and liquid phases. It characterizes the hydrophobicity of the controlled-release films by measuring their WCA using an instrument at ambient temperature. The average WCA values were obtained through measuring at five different positions on the same sample with 4 μL water droplets.

2.2.7. Measurement of the Mechanical Properties of Films

The mechanical properties of the films were evaluated using a vertical computer servo material testing machine in accordance with GB/T 1040-2018, with five replicates. The films were shaped like dumbbells and their thickness was measured using a spiral micrometer. The initial gauge length was set at 25 mm, and the measuring speed was maintained at 300 mm/min. EB and TS values were determined based on five independent drawing experiments performed under identical conditions.

2.2.8. Statistical Analyses

The data processing and statistical analyses were conducted using Microsoft Excel 2013 (Microsoft Corporation, Redmond, WA, USA). Figures were generated using Origin2018 (Origin Lab, Northampton, MA, USA). Pearson correlation analysis was employed to examine the respective relationships between the N release periods of CRFs and the WCA, WA, EB, and TS of the coating materials.

3. Results

3.1. The N Release Characteristics of CRFs

The N release characteristics of 11 types of CRFs and their N cumulative release curves are shown in Table 2 and Figure 2, respectively.

PECF exhibited η_{t1} and $\eta_{\Delta t}$ values of 0.6% and 0.4%, respectively, indicating a N release period of approximately 178 days; PPGCF coated with polypropylene glycol also displayed a slower N release characteristic, with an N cumulative release rate of 41.0% at day 28 and a N release period of around 54 days. Both PECF and PPGCF met the CRF (ISO-18644) release criteria, and their N release curves were nearly linear within the first 28 days, demonstrating good controllability. These findings suggest their widespread use, as primary products currently available on the market, in agriculture.

The two types, SALCF1/SALCF2, exhibited favorable N controlled-release properties, albeit with slightly different N release periods. Figure 2 illustrates the N release curve of SALCF2, which follows an inverse "L" shape characterized by rapid initial N release followed by a gradual decline, with a total N release period of approximately 34 days. On the other hand, SALCF1 demonstrated an N release rate equivalent to 0.2% for η_{t1} and 1.6% for $\eta_{\Delta t}$, resulting in a cumulative N release rate of 44.1% over a span of 28 days and an overall N release period approaching 50 days, thereby satisfying the nutrient demands of agricultural crops. The N release periods of different PUCFs prepared with various polyols as soft segments exhibited significant variations, as illustrated in Figure 2. PUCF formulated with PPG as a soft segment demonstrated a N release period of 54 days. Conversely, the three types of PUCFs incorporating low-molecular-weight polyols (PTMG/PEG/PCL1) as soft segments displayed comparable N release periods ranging from approximately 26 to 28 days. However, when employing PCL2 and PBA with an MW of 1000 as soft segments, the resulting η_1 values for the respective PUCFs were determined to be 55.1% and 27.8%. Furthermore, during a span of four days, the cumulative N release rates reached an alarming level of 78.2% and 78.5%, respectively, indicating that these two types of PUCFs exhibit uncontrollable behavior.

In addition, the PVACF exhibited a rapid trend of N release with a complete period of approximately 1.5 days for N release. The η_1 value of the SCF was determined to be 17.7%, followed by a significantly slower release rate thereafter. Two distinct leaching mechanisms were observed for the SCF in water, characterized by "all or nothing" performances [46]. Firstly, certain portions of the SCF surface easily crumbled and formed large pores, facilitating the quick dissolution of the fertilizer core and transforming the SCF into an empty shell; this phenomenon accounted for the initially high N release rate observed for the SCF. Secondly, other parts of the SCF remained intact even after one month's immersion in water, indicating no subsequent release.

Table 2. N release parameters of CRFs.

Coating Materials	Treatment	1 d	4 d	7 d	10 d	14 d	21 d	28 d	$\eta_{\triangle t}$	T
PVA	PVACF	68.4	98.1	99.0	99.0	99.0	99.0	99.0	--	1.5
SAL	SALCF1	0.2	1.0	2.4	4.2	7.4	18.9	44.1	1.6	50
	SALCF2	3.1	10.8	11.9	20.5	31.4	49.5	65.4	2.3	34
PE	PECF	0.6	1.5	2.2	3.3	5.4	9.7	12.8	0.4	178
PU	PPGCF	1.0	3.6	7.6	12.1	18.3	29.4	41.0	1.5	54
	PTMGCF	1.2	9.0	23.3	43.8	56.8	72.3	79.3	2.9	28
	PBACF	27.8	78.5	95.3	99.0	99.0	99.0	99.0	16.9	4
	PEGCF	6.9	40.3	63.7	75.8	80.2	84.5	87.5	3.0	25
	PCLCF1	3.7	17.8	34.2	52.3	64.2	80.4	87.1	3.1	26
	PCLCF2	55.1	78.2	89.9	93.1	95.2	95.2	95.2	7.7	4
S	SCF	17.7	20.7	21.5	22.1	22.6	23.0	23.4	--	--

Note: $\eta_{\triangle t}$ is the average release rate and T is the N release period.

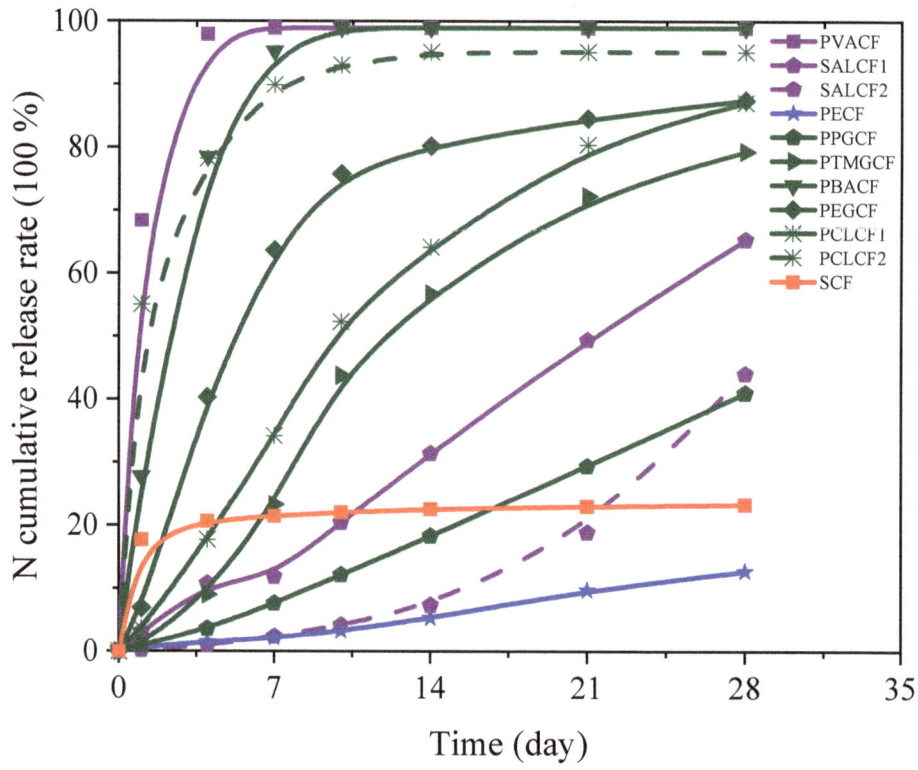

Figure 2. N cumulative release curves of different CRFs.

3.2. The Hydrophobicity of Different Coating Materials

In order to investigate the relationship between the release performance of the CRFs and the properties of the materials, flat films with different formulations were prepared and the WA and WCA of the films were measured (Figure 3a,b). The WA of different flat films varied significantly from 104.0% for PVA to 0.2% for PE. The highest WA of PVA at ambient temperature could be associated with the presence of a large number of hydroxyl groups in the PVA molecular chain. SAL1, which is a water-based material, also had a WA of 9.4%. The WAs of different PU films with different soft sections ranged from 1 to 6%. The PE film was hydrophobic in water, and the WA of the S film was similar to that of the PE film. The WCA of these films ranged from 36.1 to 92.9°. Briefly, the smallest WCA for the PVA film was 36.1° and the smallest for the SAL1 film was 66.5°. The WCAs of PU films with different soft sections were in the range of 53.5 to 83.4°. The PE film showed the highest WCA, and the S film had a similar WCA value to the PE, of 92.1°.

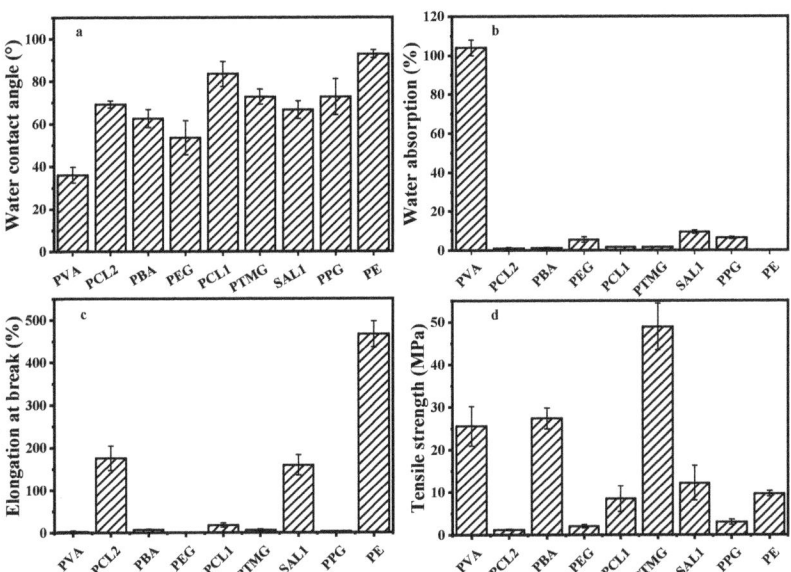

Figure 3. The hydrophobicity and mechanical properties of different coating materials. The hydrophobic properties for different coating materials are presented in (**a**) (WCA), (**b**) (WA), and the mechanical properties for different coating materials are presented in (**c**) (EB), (**d**) (TS).

The relationship between the WCA, WA, and hydrophobicity is illustrated in Figure 3. A higher WCA corresponds to a lower WA and stronger hydrophobicity. For instance, the PE film exhibited a large WCA of 92.9° and an extremely low WA of 0.04%, indicating its excellent controlled-release properties. In contrast, the PVA film had a significantly smaller WCA of 36.1° and a remarkably high WA of 104.0%, suggesting poor water resistance and uncontrolled behavior. PU films with different soft sections generally exhibited higher WCAs but lower WAs. The WCA of the PCL1 film was 83.4°; however, its WA was low, at 1.7%, while the PEG film showed a decreased WCA of 53.5° and an elevated WA of 5.5% compared to the PCL1 film. All these results suggest close mutual restraint between the hydrophobicity of the materials and the nutrient release period of CRFs.

3.3. Mechanical Properties of Different Coating Materials

The results of the EB and TS for different coating materials are presented in Figure 3c,d, suggesting the significant influence of material type on the mechanical properties of the composite films.

The three coating materials, namely, PCL2, SAL1, and PE, exhibited favorable controlled-release performances, with EB values exceeding 30%. Notably, materials with a TS ranging from 9.7 to 12.2 MPa, such as SAL1 and PE, demonstrated superior controlled-release capabilities. Among these materials, the PE film stood out due to its exceptional mechanical properties, characterized by an impressive EB value of 466.8% and a TS value of 9.7 MPa.

Additionally, the SAL1 film exhibited a relative reduction in EB (159%) and an increase in TS (12.2 MPa), resulting in a slightly stiffer and more brittle texture compared to the

PE film. Consequently, its mechanical properties and controlled-release performance also showed a corresponding decline. However, the PCL2 film with a higher EB showed inadequate strength, resulting in a suboptimal controlled-release performance.

Poor controllability was observed in almost all of the films with an EB <30%, including PVA, PBA, PEG, PCL1, PTMG, and PPG. Notably, the PVA and PBA films, with TS values of 25.6 and 27.4 MPa, respectively, exhibited a hard and brittle nature without any significant control. Conversely, the PEG and PPG films demonstrated excellent controlled-release performances, with TS values of 2 and 3 MPa, respectively. Similarly, the PTMG film displayed good controllability, with a TS value of 48.9 MPa. In conclusion, no definite relationship between the TS and EB could be established; however, materials exhibiting moderate stiffness combined with high flexibility are more suitable for use in coating applications.

3.4. Correlations between the Controlled-Release Performance of CRFs and the Physical Properties of the Coating Materials

In order to investigate the relationship between the controlled-release performance of CRFs and the hydrophobic and mechanical properties of the coating materials, we analyzed the correlations between the N release periods of CRFs and the WCA, WA, EB, and TS of the coating materials (Figure 4).

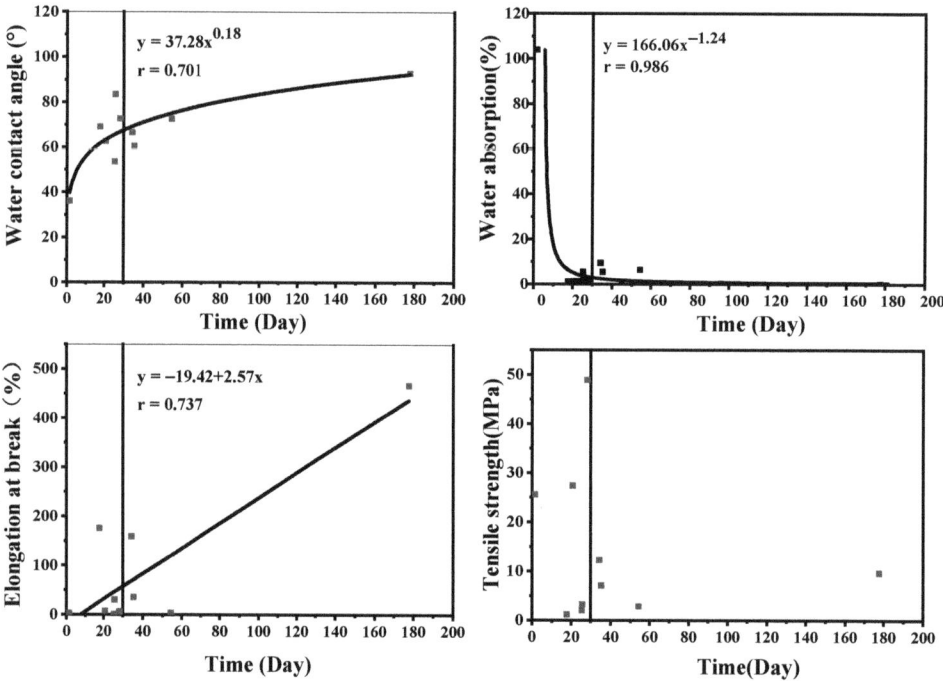

Figure 4. Correlation analyses of the N release periods of CRFs with the physical properties of their coating materials. Note: the squares in the figure are the coating materials with different release periods, and the axis of X = 30 is represent the release period of 30 days.

Figure 4 demonstrates that the correlation between the N release periods of CRFs and the WCA and WA values of coating materials can be modeled using a power function with equations $y = 37.28x^{0.18}$ and $y = 166.06x^{-1.24}$, yielding correlation coefficients of 0.701 and 0.986, respectively. The relationship between the N release periods of CRFs and the EB values of films can be modeled using a linear equation ($y = -19.42 + 2.57x$) with a correlation coefficient of 0.737; furthermore, no significant correlation was observed between the N release periods and TS values for the films tested in this study. Therefore, we selected EB as the characteristic parameter representing the mechanical properties of these films.

Based on the functional equations presented in Figure 4, the values of the material parameters required in order to obtain films with N release periods of either 30 or 90 days can be calculated and are shown in Table 3. The results indicate that for a CRF with an N release period of 30 days, coating materials should have a WA value of <2.4%, WCA value of >68.8°, and EB value of >57.7%. Specifically, when the WA is less than 0.6%, the WCA is greater than 83.8°, and the EB is increased to 211.9%, it becomes possible to produce a CRF with excellent controlled-release properties.

Table 3. Parameters of the hydrophobicity and mechanical properties of coating materials.

	Kinetic Equation	r	Release Period (Day)			
			30		90	
			y_1	y_2	y_1	y_2
WA (%)	$y = 166.06x^{-1.24}$	0.986	2.4	3.0	0.6	1.0
WCA (°)	$y = 37.28x^{0.18}$	0.701	68.8	60.0	83.8	80.0
EB (%)	$y = -19.42 + 2.57x$	0.737	57.7	30.0	211.9	100.0

Note: the calculated values were obtained by substituing the N release periods into the kinetic equation; x is the N release period and y is the parameter. y_1 is the calculated value and y_2 is the recommended value.

The results of this study demonstrate the theoretical feasibility of developing a CRF that aligns with the required nutrient release period, provided that the physical properties of the films meet the calculated values. Considering the potential limitations encountered in calculating this value due to objective factors (such as the limited availability of CRFs for factory production), we propose a recommended value (Table 3) based on the fitting results and actual development scenarios for CRFs. Consequently, we conclude that maintaining a WA below 3.0%, a WCA above 60.0°, and an EB above 30.0% is essential for ensuring the film's compliance with controlled-release requirements.

The results presented in Tables 2 and 3, as well as Figure 4, demonstrate the significant influence of the mechanical properties of coating materials on their corresponding N release period. In this study, both the PCL1 and PTMG films exhibited WCAs close to the calculated value required for producing a CRF with a N release period of 60 days. However, their actual N release periods were approximately 30 days due to their lower EBs compared to those calculated for a CRF with a N release period of 30 days. Moreover, the PTMG film showed an extended N release period compared to PCL1 due to its superior mechanical properties. Despite having lower hydrophilicity than the PCL1 and PTMG films, the SAL1 film demonstrated higher N release periods in SALCF1 formulations, exceeding 30 days due to its exceptional mechanical properties.

The results of this study demonstrate the theoretical feasibility of developing a CRF that aligns with the required nutrient release period, provided that the physical properties of the films meet the calculated value. Considering the potential limitations in calculating this value due to objective factors (such as the limited availability of CRFs for factory production), we propose a recommended value (Table 3) based on the fitting results and actual development scenarios for CRFs. Consequently, we conclude that maintaining a WA below 3.0%, a WCA above 60.0°, and an EB above 30.0% is essential for ensuring film compliance with controlled-release requirements.

3.5. The Controlled-Release Performance and Physical Properties of MSCFs

In order to validate the influence of hydrophobicity and mechanical properties on the controlled-release performance of CRFs, we modified the S material, which exhibited deficiencies in both its hydrophobicity and mechanical properties. Consequently, corresponding CRFs were produced. The properties of both the modified and unmodified films are presented in Table 4, while Figure 5 illustrates the cumulative release curves for N.

Table 4. The parameters of physical properties for different modified and unmodified films and the N release period of CRFs.

Treatment	WA (%)	WCA (°)	EB (%)	TS (MPa)	Release Period (Day)
SCF	0.1	92.1	--	--	--
MSCF1	2.4	63.3	48.3	2.8	40
MSCF2	2.2	75.9	10.4	1.2	13

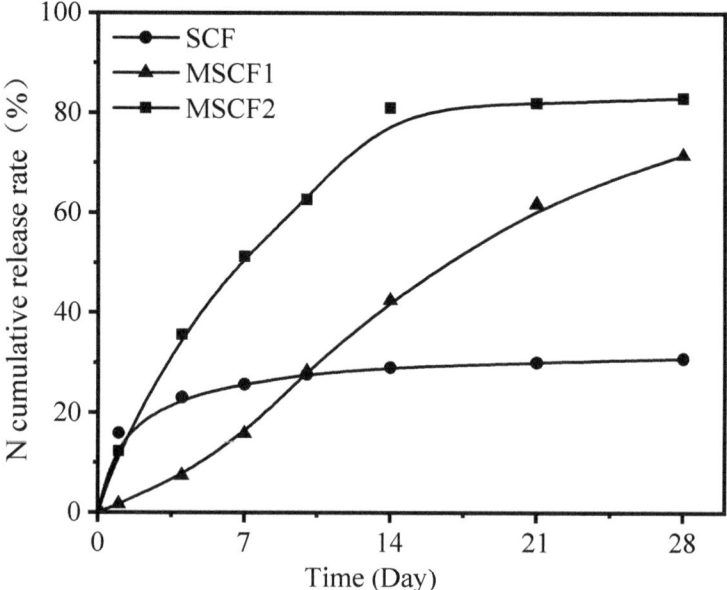

Figure 5. N cumulative release curves of unmodified and modified CRFs.

The release curve of the SCF in water, as depicted in Figure 5, exhibited a logarithmic trend with η_{t1} up to 15%. However, after 4 days at a releasing rate of 25%, it became challenging for the SCF to continue releasing nutrients, indicating an uncontrolled performance. In terms of material properties, the SCF film demonstrated high hydrophobicity (with a WCA of 92.1° and WA of 0.1%), but low TS and EB values, which were nearly negligible. When cracks appeared on the shell, a rapid infiltration of water occurred, leading to rupturing and the complete release of nutrients from individual particles. Conversely, when the shell remained intact, no infiltration occurred, and individual particles remained in perfect conditions without nutrient release. By combining S with PCL1-based PU, a soft S/PU hybrid film was obtained with increased WA values approaching the recommended levels while reducing the WCA values accordingly. However, the incorporation of PU resulted in an increase in the EB from 0 to either 10.4 or 48.3%, depending on the amount added; correspondingly, the TS also increased from 0 to either 1.2 or 2.8 MPa, respectively, thus significantly improving film flexibility upon the addition of PU content. When maintaining a ratio of S to PU at approximately equal proportions (i.e., at a ratio of approximately 1:1), both EB and TS reached their recommended values while achieving an extended N release period for MSCF1 that reached up to 40 days, thereby demonstrating good controllability for the S/PU film system, as shown in Figure 5.

4. Discussion

According to the assumptions of previously reported controlled-release models of CRFs [16,19], it is essential for the film to possess both a homogeneous structure and stable performance. It is widely recognized that the film's structure is determined by the production process, while its stability relies on the molecular properties, hydrophobicity, and mechanical stability of the coating materials. Our research has confirmed that coating materials exhibiting an excellent controlled-release performance should demonstrate adequate hydrophobicity [47] and mechanical properties simultaneously.

Considering the design of controlled-release films, ideal coating materials should fulfill three essential conditions to meet the crop nutrient release requirements: (a) insolubility in water; (b) excellent hydrophobicity; and (c) adequate mechanical stability. Based on the correlation between the film properties and nutrient release performance evaluated in this study, Figure 6 illustrates three types of controlled-release films immersed in water.

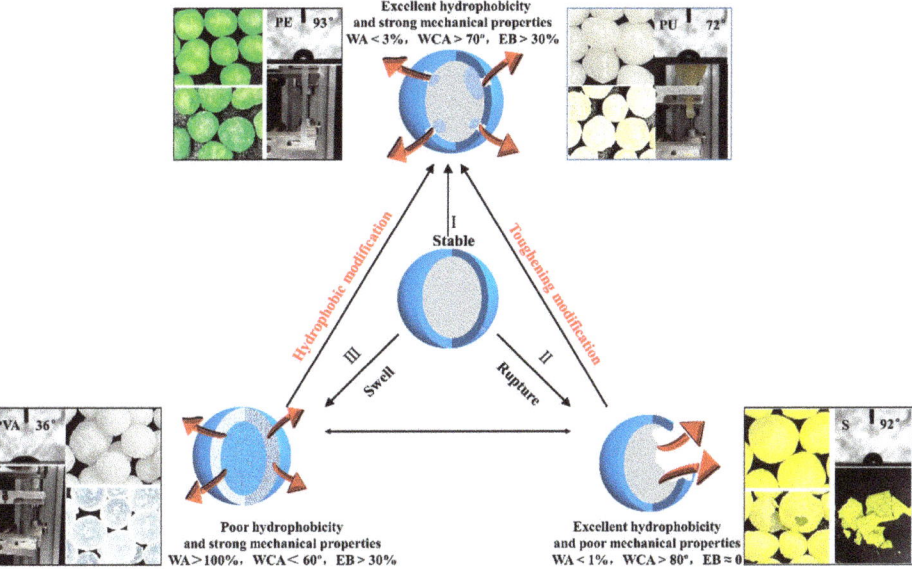

Figure 6. Schematic of the nutrient release mechanism of CRFs as influenced by their coating's material properties.

Type I: stable. When the WA is <3%, the WCA is >70°, and the EB is >30%, the film exhibits enhanced hydrophobicity and mechanical properties, simultaneously meeting conditions a, b, and c. Consequently, water permeability through the film becomes significantly hindered [18], while its pores serve as channels for nutrient release [48]. The rate of nutrient release is primarily determined by larger pores [16].

Type II: rupture. When the WA is <1%, the WCA is >80°, and the EB is ≈0, the film exhibits hydrophobic and nonabsorbent properties, thereby satisfying conditions a and b; however, it demonstrates inadequate mechanical characteristics. In its intact state, the film effectively segregates water from the nutrient, impeding the release of the core nutrient. Conversely, when the film is defective, it becomes susceptible to cracking and facilitates the rapid dissolution and release of the nutrient [49].

Type III: swell. When the WA is <100%, the WCA is <60°, and the EB is >30%, the film exhibits hydrophilicity and undergoes swelling in water [50], leading to a compromised barrier capacity within a short period, thereby demonstrating inadequate performance in terms of controlled release [51].

Coating materials matching type I primarily consist of organic synthetic polymers, such as PE and PU films. Meanwhile, certain materials matching type II are predominantly composed of inorganic coated substances, like S film. On the other hand, materials fitting into type III mainly encompass highly water-soluble polymers such as PVA film. In terms of practical applications, SCF is exclusively employed as a slow-release fertilizer. Although water-soluble polymers [51,52] and natural biomaterials [53] are environmentally friendly options, they possess high WA values and fail to meet the actual demands of agricultural production when used for coating CRFs.

At present, the predominant commercially available CRFs are primarily coated with PE or PU polymers. As agricultural chemical production involves the extensive use of polymer-coated CRFs in farmland applications, the quantity involved is substantial, resulting in higher prices. For instance, MDI costs approximately CNY 20,000/ton and polyol costs around CNY 15,000/ton. The addition of PU during CRF production with a coating rate of 4% incurs a cost of CNY 1400/ton. To address this issue, researchers have been exploring various approaches to reduce the cost of polymer-coated fertilizers by minimizing coating rates and optimizing film structures [54]. In comparison to organic materials, inorganic materials offer unparalleled price advantages. S, for example, is priced at only about CNY 2000/ton, and adding S during CRF production with a coating rate of 10% amounts to approximately CNY 200/ton—one-tenth of the price of PU. Therefore, we attempted to establish recommended values for film materials' properties to guide modifications aimed at rectifying deficiencies while preserving the favorable ecological characteristics inherent in inorganic materials. The results pertaining to the physical properties of modified-S-based materials demonstrate that aligning their material properties with the recommended values proposed herein can enhance the release performance of CRFs. Furthermore, our experiment substantiated that modifying coating materials to achieve these recommended values serves as a fundamental guarantee for producing CRFs exhibiting excellent controlled-release performance.

The modification of vegetable-oil-based PUCFs using modified bentonite resulted in a reduction in WA from 1.3 to 0.3%, an increase in WCA from 95.4 to 97.5°, and an improvement in EB from 12.3 to 20.7%. Consequently, the N release period increased from 23 to 74 days [55]. It is evident that the modified vegetable-oil-based PUCFs meet the recommended values for WA and WCA suitable for a N release period of 90 days; however, the EB falls short of our recommended value, resulting in a shorter N release period than desired (less than 90 days). If the EB can be enhanced to reach 100%, it would extend the N release period beyond 90 days. Therefore, obtaining these recommended values provides clear guidance for selecting and modifying appropriate inorganic materials or biodegradables, which serves as crucial reference information for future research work or industrialization efforts related to CRFs.

5. Conclusions

This study has investigated the relationship between the hydrophobicity and mechanical properties of coating materials and the N release periods of commercially available CRFs. The results show significant differences in the N release characteristics of CRFs coated with different materials. Longer N release periods are associated with lower WA, larger WCA, and higher EB values in the corresponding coating material. The correlation coefficients between the N release period and the WA, WCA, and EB were 0.986, 0.701, and 0.737, respectively. Based on the simulated equations, the physical properties for a coating material that can meet the release criteria include a WA less than 2.4%, a WCA greater than 68.8°, and an EB greater than 57.7%. For CRFs demonstrating a N release period of up to 30 days, the recommended values for the coating materials were WA <3.0%, WCA >60.0°, and EB >30.0%. Improving the hydrophobicity and mechanical properties of coating materials according to these recommendations could significantly enhance their controlled-release performance.

Author Contributions: Ideas and writing—original draft preparation, Y.W.; funding acquisition and writing—reviewing and editing, J.L.; formal analysis and writing—review and editing, R.L.; data collection, D.G.; writing—original draft preparation, Y.Z.; drawing, H.L.; conceptualization, ideas, supervision, project administration, and funding acquisition, X.Y. All authors have read and agreed to the published version of the manuscript.

Funding: This research was funded by the National Key R&D Program of China (2022YFD1700601), the National Natural Science Foundation of China (32372819, 31872177), and the Agricultural Science and Technology Innovation Program (ASTIP).

Data Availability Statement: The data presented in this study are available in the article.

Conflicts of Interest: The authors declare no conflict of interest.

References

1. Erisman, J.W.; Sutton, M.A.; Galloway, J.; Klimont, Z.; Winiwarter, W. How a century of ammonia synthesis changed the world. *Nat. Geosci.* **2008**, *1*, 636–639. [CrossRef]
2. Sui, B.; Feng, X.F.; Tian, G.L.; Hu, X.Y.; Shen, Q.R.; Guo, S.W. Optimizing nitrogen supply increases rice yield and nitrogen use efficiency by regulating yield formation factors. *Field Crops Res.* **2013**, *150*, 99–107. [CrossRef]
3. Zhang, X.; Davidson, E.A.; Mauzerall, D.L.; Searchinger, T.D.; Dumas, P.; Shen, Y. Managing nitrogen for sustainable development. *Nature* **2015**, *528*, 51–59. [CrossRef] [PubMed]
4. Hou, P.F.; Jiang, Y.; Yan, L.; Petropoulos, E.; Wang, J.Y.; Xue, L.H.; Yang, L.Z.; Chen, D.L. Effect of fertilization on nitrogen losses through surface runoffs in Chinese farmlands: A meta-analysis. *Sci. Total Environ.* **2021**, *793*, 148554. [CrossRef]
5. Savci, S. Investigation of effect of chemical fertilizers on environment. *APCBEE Procedia* **2012**, *1*, 287–292. [CrossRef]
6. Azeem, B.; KuShaari, K.; Man, Z.B.; Basit, A.; Thanh, T.H. Review on materials & methods to produce controlled release coated urea fertilizer. *J. Control. Release* **2014**, *181*, 11–21. [CrossRef]
7. Chen, Z.M.; Wang, Q.; Ma, J.W.; Zou, P.; Jiang, L.N. Impact of controlled-release urea on rice yield, nitrogen use efficiency and soil fertility in a single rice cropping system. *Sci. Rep.* **2020**, *10*, 10432. [CrossRef] [PubMed]
8. Xie, Y.; Tang, L.; Han, Y.L.; Yang, L.; Xie, G.X.; Peng, J.W.; Tian, C.; Zhou, X.; Liu, Q.; Rong, X.M.; et al. Reduction in nitrogen fertilizer applications by the use of polymer-coated urea: Effect on maize yields and environmental impacts of nitrogen losses. *J. Sci. Food Agric.* **2019**, *99*, 2259–2266. [CrossRef]
9. Qiao, D.L.; Liu, H.S.; Yu, L.; Bao, X.Y.; Simonb, G.P.; Petinakis, E.; Chen, L. Preparation and characterization of slow-release fertilizer encapsulated by starch-based superabsorbent polymer. *Carbohydr. Polym.* **2016**, *147*, 146–154. [CrossRef]
10. Boller, R.A.; Graver, R.B. Two-Package Coating System Comprising a Polyester Having an Acid Number of at Least 50 in One of the Packages Thereof. U.S. Patent 3,218,274, 16 November 1961.
11. Stansbury, R.L.; Lynch, C.S.; Kamil, S.; Linden, N.J. Slow-Release Fertilizer Composition Consisting of Asphalt Wax Binder and Inert Filler. U.S. Patent 3,276,857, 5 October 1964.
12. Fujita, T.; Takahashi, C.; Ohshima, M.; Ushioda, T.; Shimizu, H. Method for Producing Granular Coated Fertilizer. U.S. Patent 4,019,890, 3 December 1974.
13. Geiger, A.J.; Stelmack, E.G.; Babiak, N.M. Controlled Release Fertilizer and Method for Production Thereof. U.S. Patent 6,663,686 B1, 16 December 2000.
14. Liu, Y.H.; Wang, T.J.; Kan, C.Y.; Wang, M.H.; Jin, Y. Development in polymer latex coated fertilizer for controlled release. *Chem. Indus. Engin. Prog.* **2009**, *28*, 1589–1595. (In Chinese)
15. Lawrencia, D.; Wong, S.K.; Goh, B.H.; Goh, J.K.; Ruktanonchai, U.R.; Soottitantawat, A.; Lee, L.H.; Tang, S.Y. Controlled release fertilizers: A review on coating materials and mechanism of release. *Plants* **2021**, *10*, 238. [CrossRef]
16. Wang, Y.J.; Li, J.; Yang, X.D. The diffusion model of nutrient release from membrane pore of controlled release fertilizer. *Environ. Technol. Innov.* **2022**, *25*, 102256. [CrossRef]
17. Yang, X.D.; Jiang, R.F.; Lin, Y.Z.; Li, Y.T.; Li, J.; Zhao, B.Q. Nitrogen release characteristics of polyethylene-coated controlled-release fertilizers and their dependence on membrane pore structure. *Particuology* **2018**, *36*, 158–164. [CrossRef]
18. Lu, P.F.; Zhang, M.; Li, Q.; Xu, Y. Structure and Properties of Controlled Release Fertilizers Coated with Thermosetting Resin. *Polym.-Plast. Technol. Eng.* **2013**, *52*, 381–386. [CrossRef]
19. Shaviv, A.; Raban, S.; Zaidel, E. Modeling controlled nutrient release from a population of polymer Coated fertilizers: Statistically based model for diffusion release. *Environ. Sci. Technol.* **2003**, *37*, 2257–2261. [CrossRef] [PubMed]
20. Trinh, T.H.; KuShaari, K.; Basit, A. Modeling the release of nitrogen from controlled release fertilizer with imperfect coating in soils and water. *Ind. Eng. Chem. Res.* **2015**, *54*, 6724–6733. [CrossRef]
21. Cruz, D.; Bortoletto-Santos, R.; Guimarães, G.; Polito, W.L.; Ribeiro, C. Role of polymeric coating on the phosphate availability as a fertilizer: Insight from phosphate release by castor polyurethane coatings. *J. Agric. Food Chem.* **2017**, *65*, 5890–5895. [CrossRef]
22. Du, C.W.; Zhou, J.; Shaviv, A.; Wang, H. Mathematical model for potassium release from polymer-coated fertilizer. *Biosyst. Eng.* **2004**, *88*, 395–400. [CrossRef]
23. Wang, G.D.; Yang, L.; Lan, R.; Wang, T.J.; Jin, Y. Granulation by spray coating aqueous solution of ammonium sulfate to produce large spherical granules in a fluidized bed. *Particuology* **2013**, *11*, 483–489. [CrossRef]
24. Lan, R.; Wang, G.D.; Yang, L.; Wang, T.J.; Kan, C.Y.; Jin, Y. Prediction of release characteristics of film-coated urea from structure characterization data of the film. *Chem. Eng. Technol.* **2013**, *36*, 347–354. [CrossRef]

25. Yang, L.; An, D.; Wang, T.J.; Kan, C.Y.; Jin, Y. A model for the swelling of and diffusion from a hydrophilic film for controlled-release urea particles. *Particuology* **2017**, *30*, 73–82. [CrossRef]
26. Ariyanti, S.; Man, Z.; Bustam, M.A. Improvement of hydrophobicity of urea modified tapioca starch film with lignin for slow release fertilizer. *Adv. Mater. Res.* **2020**, *626*, 350–354. [CrossRef]
27. Shen, Y.Z.; Zhou, J.M.; Du, C.W.; Zhou, Z.J. Hydrophobic modification of waterborne polymer slows urea release and improves nitrogen use efficiency in rice. *Sci. Total Environ.* **2021**, *794*, 148612. [CrossRef] [PubMed]
28. Treinyte, J.; Grazuleviciene, V.; Paleckiene, R.; Ostrauskaite, J.; Cesoniene, L. Biodegradable polymer composites as coating materials for granular fertilizers. *J. Polym. Environ.* **2018**, *26*, 543–554. [CrossRef]
29. Han, X.Z.; Chen, S.S.; Hu, X.G. Controlled-release fertilizer encapsulated by starch/polyvinyl alcohol coating. *Desalination* **2009**, *240*, 21–26. [CrossRef]
30. Chen, S.L.; Jiang, Y.F.; Chang, B.; Yang, M.; Zou, H.T.; Zhang, Y.L. Preparation and characteristics of urea coated with water-based copolymer-biochar composite film material. *J. Plant Nutr. Fert. Sci.* **2018**, *24*, 1245–1254. (In Chinese) [CrossRef]
31. Sarkar, A.; Biswas, D.R.; Datta, S.C.; Dwivedi, B.S.; Bhattacharyya, R.; Kumar, R.; Bandyopadhyay, K.K.; Saha, M.; Chawla, G.; Saha, J.K.; et al. Preparation of novel biodegradable starch/poly (vinyl alcohol)/bentonite grafted polymeric films for fertilizer encapsulation. *Carbohydr. Polym.* **2021**, *259*, 117679. [CrossRef]
32. Bai, Y.; Chen, S.L.; Fan, L.J.; Yang, M.; Zou, H.T.; Zhang, Y.L. Preparation and properties of nano-SiO_2–polyvinyl alcohol–γ-polyglutamic acid composite film materials. *J. Plant Nutr. Fert. Sci.* **2019**, *25*, 2044–2052. (In Chinese) [CrossRef]
33. Yang, Y.; Zou, H.T.; Wang, J.; Xu, M.; Liu, Y.; Zhang, Y.L. Preparation and properties of modified polyvinyl alcohol film for encapsulation of fertilizer. *J. Plant Nutr. Fert. Sci.* **2012**, *18*, 1286–1292. (In Chinese)
34. Roach, P.; Shirtcliffe, N.J.; Newton, M.I. Progess in superhydrophobic surface development. *Soft Matter.* **2008**, *4*, 24–40. [CrossRef]
35. Xie, J.Z.; Yang, Y.C.; Gao, B.; Wan, Y.S.; Li, Y.C.C.; Xu, J.; Zhao, Q.H. Biomimetic superhydrophobic biobased polyurethane-coated fertilizer with atmosphere "outerwear". *ACS Appl. Mater. Interfaces* **2017**, *9*, 15868–15879. [CrossRef]
36. Zhang, S.G.; Yang, Y.C.; Gao, B.; Li, Y.C.C.; Liu, Z.G. Superhydrophobic controlled-release fertilizers coated with bio-based polymers with organosilicon and nano-silica modifications. *J. Mater. Chem. A* **2017**, *5*, 19943–19953. [CrossRef]
37. Yang, X.D.; Zhao, B.Q.; Li, Y.T.; Li, J.; Lin, Z.A.; Yuan, L. Method for Producing Controlled-Release Fertilizer Coated with Polyurethane. U.S. Patent 9,416,064 B2, 16 August 2014.
38. Li, L.X.; Sun, Y.M.; Cao, B.; Song, H.H.; Xiao, Q.; Yi, W.P. Preparation and performance of polyurethane/mesoporous silica composites for coated urea. *Mater. Des.* **2016**, *99*, 21–25. [CrossRef]
39. Dai, C.; Yang, L.; Xie, J.R.; Wang, T.J. Nutrient diffusion control of fertilizer granules coated with a gradient hydrophobic film. *Colloids Surf. A* **2020**, *588*, 124361. [CrossRef]
40. Kassem, I.; Ablouh, E.H.; Bouchtaoui, F.Z.E.; Kassab, Z.; Khouloud, M.; Sehaqui, H.; Ghalfi, H.; Alami, J.; Achaby, M.E. Cellulose nanocrystals-filled poly (vinyl alcohol) nanocomposites as waterborne coating materials of NPK fertilizer with slow release and water retention properties. *Int. J. Biol. Macromol.* **2021**, *189*, 1029–1042. [CrossRef] [PubMed]
41. Liu, Y.H.; Wang, T.J.; Qin, L.; Jin, Y. Urea particle coating for controlled release by using DCPD modified sulfur. *Powder Technol.* **2008**, *183*, 88–93. [CrossRef]
42. Chen, Z.; Yang, X.D.; Wang, N.; Zhang, X.; Zhang, J.; Jiang, Z.G. Prepartion and characterization of modified SiO_2/PU composite coating for controlled- release fertilizer. *New Chem. Mater.* **2020**, *48*, 146–150. (In Chinese) [CrossRef]
43. Wang, S.P.; Li, X.; Ren, K.; Huang, R.; Lei, G.C.; Shen, L.J. Surface modification of pyrophyllite for optimizing properties of castor oil-based polyurethane composite and its application in controlled-release fertilizer. *Arab. J. Chem.* **2023**, *16*, 104400. [CrossRef]
44. Cao, Y.P.; Yang, X.D.; Jiang, R.F.; Zhang, F.S.; Hu, S.W. A Polymer Coated Controlled Release Fertilizer and Its Production Method and Special Coating Material. ZL 200710099144.7, 14 May 2007. (In Chinese).
45. *ISO 18644:2016*; Fertilizers and Soil Conditioners—Controlled-release Fertilizer: General Requirements. ISO: Geneva, Switzerland, 2016.
46. Lu, H.; Dun, C.P.; Jariwala, H.; Wang, R.; Cui, P.Y.; Zhang, H.P.; Dai, Q.G.; Yang, S.; Zhang, H.C. Improvement of bio-based polyurethane and its optimal application in controlled release fertilizer. *J. Control. Release* **2022**, *350*, 748–760. [CrossRef]
47. Tapia-Hernández, J.A.; Madera-Santana, T.J.; Rodríguez-Félix, F.; Barreras-Urbina, C.G. Controlled and prolonged release systems of urea from micro and nanomaterials as an alternative for developing a sustainable agriculture: A review. *J. Nanomater.* **2022**, *2022*, 5697803. [CrossRef]
48. Duan, L.L.; Zhang, M.; Liu, G.; Yang, Y.C.; Yang, Y. Membrane microstructures and nutrient release mechanism of thermoplastic coated urea. *J. Plant Nutr. Fert. Sci.* **2019**, *15*, 1170–1178. (In Chinese)
49. Ibrahim, K.R.M.; Babadi, F.E.; Yunus, R. Comparative performance of different urea coating materials for slow release. *Particuology* **2014**, *17*, 165–172. [CrossRef]
50. Jamnongkan, T.; Kaewpirom, S. Potassium release kinetics and water retention of controlled-release fertilizers based on chitosan hydrogels. *J. Polym. Environ.* **2010**, *18*, 413–421. [CrossRef]
51. Mulder, W.J.; Gosselink, R.J.A.; Vingerhoeds, M.H.; Harmsen, P.F.H.; Eastham, D. Lignin based controlled release coatings. *Ind. Crops Prod.* **2011**, *34*, 915–920. [CrossRef]
52. Ferna, M.; Garrido-Herrera, F.J.; Gonza, E.; Villafranca-Sa, M.; Flores-Ce, F. Lignin and ethylcellulose as polymers in controlled release formulations of urea. *J. Appl. Polym. Sci.* **2008**, *108*, 3796–3803. [CrossRef]

53. Wang, F.Y.; Liu, M.Z.; Ni, B.O.; Xie, L.H. κ-carrageenan–sodium alginate beads and superabsorbent coated nitrogen fertilizer with slow-release, water-retention, and anticompaction properties. *Ind. Eng. Chem. Res.* **2012**, *51*, 1413–1422. [CrossRef]
54. Li, L.X.; Wang, M.; Wu, X.D.; Yi, W.P.; Xiao, Q. Bio-based polyurethane nanocomposite thin coatings from two comparable POSS with eight same vertex groups for controlled release urea. *Sci. Rep.* **2021**, *11*, 9917. [CrossRef]
55. Zhao, M.H.; Wang, Y.Q.; Liu, L.X.; Liu, L.X.; Chen, M.; Zhang, C.Q.; Lu, Q.M. Green coatings from renewable modified bentonite and vegetable oil based polyurethane for slow release fertilizers. *Polym. Compos.* **2017**, *39*, 4355–4363. [CrossRef]

Disclaimer/Publisher's Note: The statements, opinions and data contained in all publications are solely those of the individual author(s) and contributor(s) and not of MDPI and/or the editor(s). MDPI and/or the editor(s) disclaim responsibility for any injury to people or property resulting from any ideas, methods, instructions or products referred to in the content.

Article

Green Preparation of Lightweight, High-Strength Cellulose-Based Foam and Evaluation of Its Adsorption Properties

Yongxing Zhou [1], Wenbo Yin [1], Yuliang Guo [2], Chenni Qin [1], Yizheng Qin [1] and Yang Liu [1,3,*]

[1] College of Light Industry and Food Engineering, Guangxi University, Nanning 530004, China; 2005170220@st.gxu.edu.cn (Y.Z.)
[2] Shandong Institute of Standardization, Jinan 250000, China
[3] Guangxi Key Laboratory of Clean Pulp and Paper and Pollution Control, Guangxi University, Nanning 530004, China
* Correspondence: xiaobai@gxu.edu.cn

Abstract: In recent years, the application scope of most cellulose-based foams is limited due to their low adsorbability and poor recyclability. In this study, a green solvent is used to extract and dissolve cellulose, and the structural stability of the solid foam is enhanced by adding a secondary liquid via the capillary foam technology, and the strength of the solid foam is improved. In addition, the effects of the addition of different gelatin concentrations on the micro-morphology, crystal structure, mechanical properties, adsorption, and recyclability of the cellulose-based foam are investigated. The results show that the cellulose-based foam structure becomes compact, the crystallinity is decreased, the disorder is increased, and the mechanical properties are improved, but its circulation capacity is decreased. When the volume fraction of gelatin is 2.4%, the mechanical properties of foam are the best. The stress of the foam is 55.746 kPa at 60% deformation, and the adsorption capacity reaches 57.061 g/g. The results can serve as a reference for the preparation of highly stable cellulose-based solid foams with excellent adsorption properties.

Keywords: cellulose-based foam; gelatin; adsorption

1. Introduction

Crude oil leakage and industrial emissions have led to high levels of pollution in the ecological environment. Recently, bio-waste and natural resources have been considered for developing green synthetic nanomaterials/nanoparticles. These green nanoparticles may be employed as a viable alternative to current methods of pollution remediation since they are cheap, stable, safe, and environmentally benign [1]. Among them, the method of adsorbing oil-water by materials with adsorption properties has been widely concerned in pollution control. Cellulose-based foam has attracted much attention as an adsorbent and is widely reported in the literature [2–5] as one of the sewage treatment materials [3,5–12] with the highest potential owing to its light weight, low density, high porosity, and high liquid holdup [13]. However, chemical modifications can often cause serious damage to the network structure of the foam and reduce its service life. Therefore, it is crucial to find an oil-absorbing cellulose-based foam with a stable structure and good recyclability performance.

Cellulose is the primary component of lignocellulosic biomass which can serve as a plentiful source of carbohydrates for the production of numerous high-demand chemicals [14]. The process of preparing cellulose-based foams requires the extraction and dissolution of cellulose by a green method. Cellulose extraction techniques include mechanical extraction by homogenization, mechanical isolation by steam explosion, defibrillation by high-intensity ultrasonication, extraction by electrospinning technique, extraction and dissolution of cellulose using ionic liquids, etc. [15]. The traditional cellulose extraction and dissolution process is not in line with the current trend of energy saving and environmental protection. Therefore, in order to fully utilize cellulose resources, it is necessary to develop "green" cellulose extraction methods and suitable cellulose dissolution pathways. As neoteric green solvents, ionic liquids (ILs) refer to a specific class of molten salts which are liquids at temperatures of 100 °C [16,17]. In recent years, many researchers have investigated the dissolution of cellulose in green solvents(ILs), Froschauer et al. [18] separated cellulose and hemicellulose from wood pulp by 1–ethyl–3–methylimidazolium acetate [C2mim][Ac]/Cosolvent (water, ethanol, or acetone) systems. Zhuo et al. [19] synthesized SO_3H^- functionalized acidic ionic liquids used as catalysts for the hydrolysis of cellulose in [C4mim][Cl] and so on. The dissolution mechanism of cellulose in ILs [20] involves the oxygen and hydrogen atoms of cellulose-OH in the formation of electron donor-electron acceptor (EDA) complexes which interact with the ionic liquid. The cations in ionic liquid solvents act as the electron acceptor center and anion as the electron-donor center. The two centers must be located close enough in space to permit the interactions and to permit the EDA complexes to form. Upon interaction of the cellulose-OH and the ionic liquid, the oxygen and hydrogen atoms from hydroxyl groups are separated, resulting in the opening of the hydrogen bonds between molecular chains of the cellulose, and finally, the cellulose dissolves [21].

Increasing the stability of wet foam is considered an effective method to prevent structural damage of the cellulose-based foam network. Foam is a complex metastable system that is influenced by many factors. According to the Reynolds equation [22], foam decay is related to the liquid film thickness, solution viscosity, and the size of bubbles [23,24]. Decay is often attributed to the foam thinning and rupture caused by the liquid film drainage, which is controlled by the viscoelasticity of the liquid film. Therefore, the viscoelasticity of the liquid film is a very important factor influencing the stability of the foam. In general, adding a surfactant will play a positive role in improving the stability of the foam. The foam forming of cellulose-based materials is influenced by the surfactant type and dosage [25]. During foam formation, the surfactants are arranged in an ordered manner on the liquid film surface, wherein the hydrophilic group is directed toward the water and the hydrophobic group is directed toward the air [26]. When the concentration of surfactants reaches the critical micelle concentration, the interface adsorption is saturated and micelles are formed. On one hand, the stabilization mechanism reduces the liquid discharge power by reducing the Laplace pressure; on the other hand, it imparts the liquid film with a certain degree of Gibbs film elasticity. When the liquid film is impacted, local deformation occurs. Consequently, the adsorption density of the surfactant in this area decreases, and a surface tension gradient is formed within the adjacent area. Therefore, the liquid in the liquid film flows from the low-surface-tension area to the high-surface-tension area [22], thereby preventing deformation (thinning) and rupture of the liquid film. Moreover, the increase in solution viscosity plays a secondary role in the stability of the foam. According to the literature [27], within the range of foaming ability, higher solution viscosity leads to a more stable foam. Chen et al. [28] studied the stability and rheological properties of foams prepared from surfactants and clay particle dispersion. The author thus confirmed that an increase in solution viscosity can indeed enhance the stability of the foam. In the case of cellulose-based wet foams, the size of the foam directly determines the pore size of the subsequent cellulose porous foam. Compared to the pure water-vapor two-phase foam, the cellulose-based wet foam is subjected to a much stronger capillary force and external force during the curing process, which renders the foam extremely unstable. Therefore, the method of formation of a dense network structure is important for the preparation of stable cellulose-based wet foams. In a previous study [29], the cationic character of a cationic polyacrylamide (CPAM) was successfully exploited to capture the negatively charged free nanofibers (NFC) in an aqueous solution. Therefore, during the preparation of a cellulose-based stable wet foam, the electrostatic force between CPAM and NFC can be fully utilized. The NFC is used as a patch, and the local adsorption of CPAM on the large cellulose is used as a bonding point to fill the NFC in the gap of the cellulose skeleton [30], and this is called the "patch bridge" mechanism. By changing the amount of NFC added, the spatial distance between the cellulose is changed, thus changing the density of the cellulose network and thereby improving that stability of the cellulose-based wet foam. Furthermore, the factors affecting the stability of the cellulose-based wet foams also include the foaming agent concentration, cellulose concentration, stirring speed, etc. [31–34].

In addition, gelatin is generally utilized for structure changes and increased porosity [35]. To improve the resistance of the cellulose network structure, a capillary foam can be formed by adding a second liquid (gelatin) to increase the yield stress, thereby reinforcing the cellulose skeleton. According to literature reports [36,37], the increase in the yield stress can be explained through two possibilities: (1) formation of physical bonds between the particles after adding the secondary liquid and (2) hydrogen bonding between the cellulose and the gelatin and the increase in the volume of the single-capillary bridge connecting the cellulose [38]. Moreover, it was found that, when an external force was applied, the stress can be transferred in the polymer network structure and dispersed [39]. Hydrogen bonds are easy to design and have dynamic reversibility under external stimulation [40]. With the addition of the second liquid, the cellulose gradually forms a network that is connected by liquid bridges across the bubbles, with a significant increase in the yield stress [37]. When the critical volume fraction is reached, the number of liquid bridges reaches saturation, indicating an insignificant increase in the yield stress. However, when the volume fraction increases beyond the critical value, the yield stress increases rapidly. At this time, the increase in the volume of capillary liquid bridges between the cellulose generates a strong yield stress [37]. The results show that the second liquid is imperative for the formation of a network of particles interconnected by liquid bridges, which then changes into a gel-like substance with high elasticity [41]. This transformation significantly increases the yield stress and viscosity, which are higher than those of an ordinary cellulose foam. Therefore, the cellulose skeleton is enhanced, and the resistance ability of the cellulose network structure is improved.

In this study, gelatin is used as the second liquid, with capillary foam technology to improve the stability of the cellulose skeleton. The effects of gelatin concentration in the foam on the microstructure, crystal structure, and mechanical properties of the cellulose-based foam are studied. Furthermore, the adsorption and cycle performance of the cellulose-based foam with different gelatin concentrations are analyzed by adsorption cycle experiments which are carried out by using the traditional extrusion–drying–absorption cycle method. However, there is still a lack of clear analysis of the adsorption cycle times, and we hope that some researchers will conduct detailed and clear research in the future.

2. Materials and Methods

2.1. Materials

Bagasse fiber (BF) was obtained from Guangxi Guitang Group. Nanofibrillated cellulose (NFC) was purchased from Tianjin Damao Chemical Reagent Factory. Sodium lauryl sulfate (SDS), 1-tetradecyl alcohol (TDA), and gelatin were purchased from Shanghai Aladdin Biochemical Technology Co., Ltd. (Shanghai, China). Gum arabic (GAC) was purchased from Tianjin Damao Chemical Reagent Factory. All reagents were used as is, unless otherwise stated.

2.2. Experimental Devices

The experimental devices used are as follows: high shear dispersion emulsifier (FM 200, Frug Fluid Machinery Co., Ltd., Foshan, China), high-speed disperser (Ultra-Turrax, Aika, Germany), thermostat water bath cauldron (DF-101S, Bonsai Instruments Co., Ltd., Shanghai, China), freeze-dryer (E35 A-Pro, Shanghai Qiaofeng Industrial Co., Ltd., Shanghai, China), field emission scanning electron microscope (Gemini500, Zeiss Instruments, Germany), X-ray computed tomography (GE Vtomx, GE, the US), automatic mercury injection apparatus (AutoPore IV 9500, McMurray, the US), X-ray diffractometer (Rigaku D/MAX

2500V, Neo-Confucianism, Japan), and universal tensile testing machine (LS1, AMETEK, the US).

2.3. Preparation of Cellulose-Based Foam

The cellulose-based composites are developed by dispersing the stable aqueous suspensions of cellulose mostly in hydrosoluble or hydrodispersible or latex-form polymers [15], and the cellulose-based foam is prepared by the freezing method. At first, the study uses SDS as a green solvent, and 0.015 g/mL SDS, 0.1 g/mL GAC, and 0.015 g/mL TDA solutions are used as foaming solutions. The solution is heated in a water bath at 60 °C for 5 min and then taken out when its appearance changes from turbid into clear light blue. The solution is cooled to room temperature and then transferred into a blast furnace with a dry weight accounting for 1.8% of the whole system. NFC is then dispersed by an emulsifier and added to 20% of BF dry weight. The mixture is stirred at a low speed of 800 rpm for 10 min and then foamed at a high speed of 2000 rpm for 15 min. After foaming, a small amount of CPAM is added dropwise while stirring constantly. After stirring for 5 min, the gelatin is added dropwise until the volume accounts for 0%, 1.2%, 1.6%, 2.4%, and 3.2% of the total weight of the system. The samples are named NPB, NPGB-1.2, NPGB-1.6, NPGB-2.4, and NPGB-3.2, respectively. Then, the solution is poured into a round mold of a 10 cm diameter and frozen overnight in a refrigerator at −20 °C. After the above series of operations, the foam is dried in a freeze dryer at −65 °C for 36 h to obtain a solid foam containing different concentrations of gelatin.

2.4. Field-Emission Scanning Electron Microscopy (FESEM)

The microstructure of the cellulose-based foam is analyzed by FESEM (Gemini500, Giebelstadt, Germany). Before testing, the foam is quenched in liquid nitrogen and adhered to the stage with a conductive adhesive. All samples are gold sprayed under vacuum for 60 s, and the samples are tested at 10 kV. The porosity is calculated according to the Formula (1): [42,43]

$$Porosity = \left(1 - \frac{\rho_b}{\rho_s}\right) \times 100\% \tag{1}$$

$$\rho_b = \frac{m}{v} \tag{2}$$

ρ_b is the bulk density of cellulose foam, ρ_s is the skeleton density of cellulose, which is 1.5 g/cm^3, and ρ_b is the density of the foam material, calculated according to Formula (2) [44].

2.5. X-ray Computed Tomography (Micro-CT)

The cellulose-based solid-state foam is cut into small pieces of 1 × 1 × 1 cm and scanned at the source voltage of 20 kV X-ray tube, with an image resolution of 2 µm.

2.6. Pore-Size Distribution

The pore-size distribution of the foam is measured by an automatic mercury injection apparatus. The cellulose-based foam is cut into pieces of 1.5 cm × 1 cm × 1 cm and dried overnight in an oven at 60 °C before the measurement. The measured contact angle is 130°, and the pressure ranged from 0.10 to 61,000 psia.

2.7. X-ray Diffraction (XRD)

The crystal structure of the foam cellulose is analyzed by XRD. The samples are cut into 1 mm thick slices for testing. Cu-Kα rays with a wavelength of λ = 0.154 nm are used for scanning analysis. The voltage is 40 kV, the current is 30 mA, scanning diffraction angle "2θ" is within the range 5–50°, and scanning speed is 3 °/min. Combined with the analysis software JADE, the crystallinity index (CrI) is calculated according to Formula (3) [45].

$$CrI = \frac{I_{002} - I_{am}}{I_{002}} \times 100\% \tag{3}$$

The diffraction peak intensity obtained when I_{002} is at $2\theta = 22.5°$ is the diffraction intensity attributed to the crystalline region, and the diffraction peak intensity obtained when I_{am} is at $2\theta = 16.5°$ is the diffraction intensity attributed to the amorphous region.

2.8. Compression Testing

The mechanical strength of the foams is analyzed using a universal tensile tester (LS1, USA). The compressive properties of cellulose-based foams are tested according to GB/T 8813-2008 "determination of compression properties of rigid foams". For this test, cylindrical specimens with a height of 15 mm and a diameter of 20 mm are prepared using a die with a height of 17 mm and a diameter of 22.5 mm. Before the test, all samples are placed in a vacuum drying oven and kept at 25 °C for 12 h. The test is carried out at 10 mm/min until the material strain reaches 80%.

2.9. Adsorption Cycle Capacity Test

Two common organic solvents and six different oils are used as adsorbents. The cellulose-based foam is cut into small pieces of 20 mm × 20 mm × 10 mm, and the initial weight of each sample is denoted by M_0. In the adsorption experiment, the sample is immersed in different adsorption solvents until reaching adsorption saturation, and then the excess surface liquid is scraped off with paper. The weight after adsorption is denoted by M_t, and the liquid adsorption Q_s is calculated as follows (4): [46]

$$Q_s = \frac{M_t - M_0}{M_0} \qquad (4)$$

The desorption step involves simply squeezing the adsorbed samples, washing them with ethanol and soaking them overnight and finally drying them in a vacuum oven at 60 °C for 10 h. The whole above process is referred to as a cycle.

3. Results and Discussion

3.1. Morphological Characteristics of the Cellulose-Based Foam

The cellulose-based foam is prepared by the foaming–molding method, and its cell structure is directly related to the strength of the network structure of the cellulose-based stable wet foam before curing. Therefore, under the conditions of different gelatin concentrations, the final cellulose-based foam has obvious differences in appearance and morphology.

Figure 1a shows an actual foam picture before (NPB) and after adding gelatin (NPGB). It can be seen from the figure that the foam of the former is softer than that of the latter and has a poorer performance, which makes it difficult to perform the relevant characterization by scanning electron microscopy. However, by comparing the X-ray diffraction patterns of the two (Figure 1b), it is found that the foam structure after adding gelatin becomes significantly denser. In the figure, black represents the pores, white represents the cellulose, and the brighter the color, the denser is the cellulose or the cellulose distribution at that location. As revealed in the figure, the white highlighted areas of the foam are banded together without gelatin, and the distance between the bands are large. This indicates a phenomenon where the cellulose gathers in piles, and the skeleton structure in the foam is hollow. With the addition of gelatin, the brightness of the white-banded area decreases, and the area itself also shrinks. It is evenly distributed in the form of bright white dots. The above phenomenon indicates that the cellulose is uniformly dispersed and its structure is compact. The results show that the high yield stress is imparted to the cellulose skeleton via the capillary foam technology. Therefore, it can successfully maintain the network structure of the foam, which can also be reflected by the appearance of the foam (Figure 1a). NPB has a soft structure, while NPGB is smooth and delicate with a firm texture.

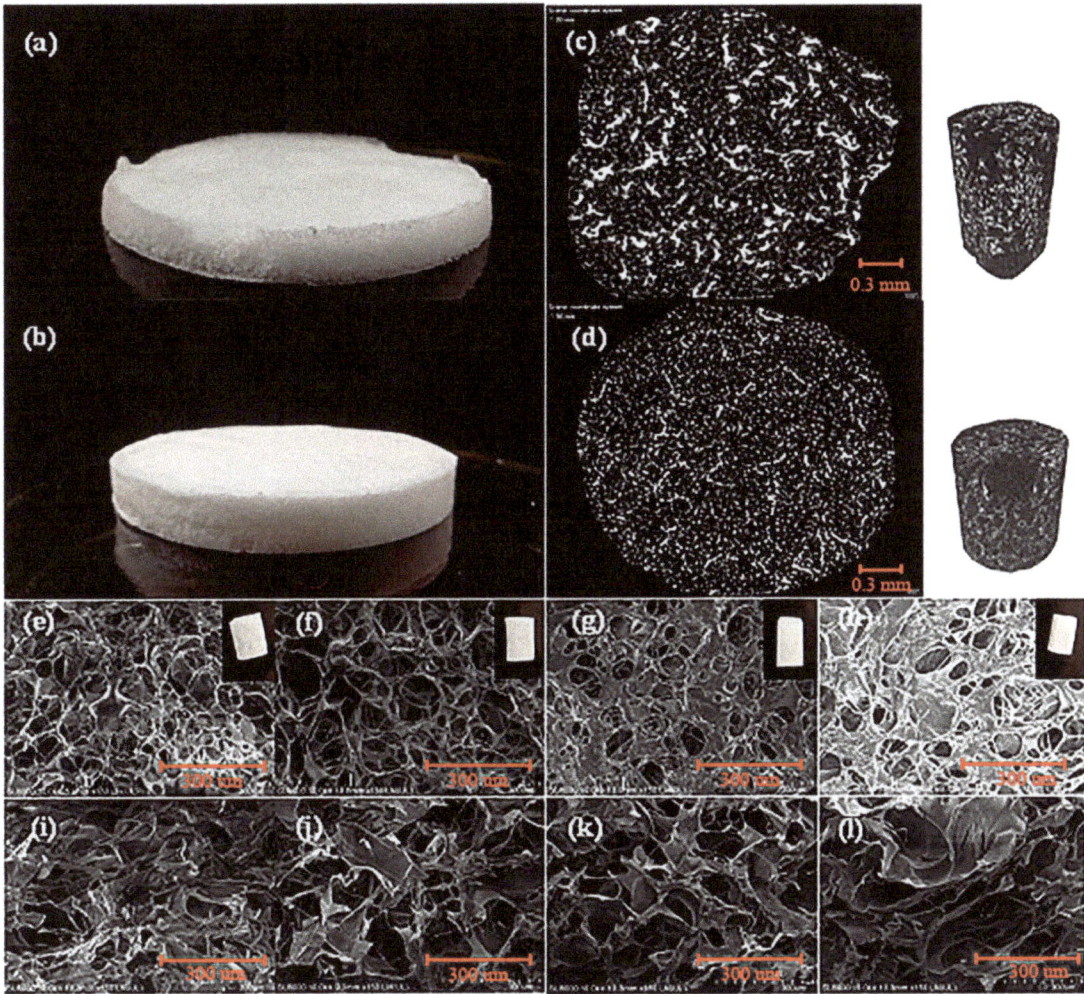

Figure 1. (**a**) Physical appearance of NPB and NPGB foams; (**b**) X-ray diffraction profile; and (**c**,**d**) surface (**e**–**h**) and section (**i**–**l**) SEM images of the cellulose-based foam at different gelatin volume fractions (from the left to the right, they are NPGB-1.2, NPGB-1.6, NPGB-2.4, and NPGB-3.2).

Figure 1c shows the surface (top) and cross-sectional (bottom) SEM images of cellulose-based foams with different gelatin concentrations. It can be seen from the surface SEM images that all the samples maintain a good cell structure in the transverse direction. However, in the machine direction, when the volume fraction of gelatin is 1.2%, the foam appears hollow, which is known as the collapse phenomenon. This is because although the gelatin provides strength to the fibrous matrix of the foam at this stage, the volume fraction is not sufficient to resist external damage. When the volume fraction of gelatin is 1.6%, there is no hollow or collapse phenomenon in the longitudinal direction, but the cellulose is messy. The results show that under these conditions, although the foam network structure can resist the damage during molding to a certain extent, the addition of gelatin is still very low. Therefore, the degree of physical crosslinking is insufficient, and the foam cannot retain not its complete network structure vertically. In contrast, when the volume fraction of gelatin is 2.4% or 3.2%, a significant change in the foam morphology can be seen in the surface SEM images. This is mainly due to the physical cross-linking between the gelatin and the cellulose. It can be seen from the cross-sectional SEM image that the same cell structure as that of the foam surface appears in the longitudinal direction, the cell wall is firm, and there is no bending phenomenon, indicating that a three-dimensional network structure is formed, which is beneficial to improve the mechanical strength of the foam. Among the samples, when the volume fraction of gelatin is as high as 3.2%, there is an excessive crosslinking in the foam. The cellulose connected by the second liquid is piled up to form lamellae; the interaction between the gelatin and the cellulose is hindered by the cross-linking effect between the gelatin and the gelatin molecules in lamellae [47]. This results in a stratification phenomenon, as shown in the figure in the longitudinal direction. Table 1 shows the density and porosity of the cellulose-based foam prepared under different gelatin concentrations, where the density is less than 0.01 g/cm^3. At the same time, the porosity reaches more than 98%, meeting the requirements of ultra-lightweight and porous.

Table 1. Physical parameters of the cellulose-based foam at different gelatin volume fractions.

Sample	Density (g/cm^3)	Porosity (%)
NPB	0.0108 ± 0.0012	99.30 ± 0.05
NPGB-1.2	0.0155 ± 0.0030	98.95 ± 0.15
NPGB-1.6	0.0145 ± 0.0050	99.03 ± 0.05
NPGB-2.4	0.0175 ± 0.0050	98.84 ± 0.03
NPGB-3.2	0.0173 ± 0.0034	98.86 ± 0.23

3.2. Pore Size Distribution of the Cellulose-Based Foam

An automatic mercury injection apparatus is used to characterize the pore size of cellulose-based foams with different gelatin concentrations. Figure 2a shows the mercury injection curve. In the low-pressure region, the mercury injection amount gradually increases, this stage mainly fills the large hole. With the increase in pressure, the amount of mercury injected is no longer increasing. At this time, the energy consumption is mainly used for the volumetric compression of the material [48]. It can be seen from the curve that the foam sample has a macroporous structure (having pores larger than 500 nm is called macroporous). When the volume fraction of gelatin is 0 or 1.2%, the constant value of the amount of mercury injected is the same. With the increase in the volume fraction of gelatin, the amount of mercury injected increases significantly. This indicates that when the volume fraction of gelatin is 0 or 1.2%, there are more large-aperture bubbles. The detailed pore size distribution is shown in Figure 2b. Without the addition of gelatin, the pore size distribution of the foam is uniform and large. After the addition of the gelatin, the pore size of the foam is significantly reduced. As the volume fraction of the gelatin increases, the pore size distribution curve of the foam exhibits a multimodal phenomenon. When the volume fraction of the gelatin is 2.4%, the one strong peak that appears in the large-size region, as in other concentrations, is accompanied by other small peaks that appear in the small-size region. This indicates the presence of a hierarchically interconnected porous structure, which is beneficial to improve the mechanical properties of the foam [49]. The detailed data are shown in Figure 2c. With the increase in the gelatin volume fraction, the pore size of the foam first decreases and then increases. When the volume fraction of the gelatin is 2.4%, the pore size of the foam is the smallest, and the average pore size is approximately 47 μm, which is 178 μm less than the pore size before adding gelatin, which was 225 μm. The results show that based on the temperature-controlling properties of gelatin, the addition of a second liquid can not only improve the strength of the cellulose skeleton but also refine the pore size of the foam.

In addition, it has been confirmed that maintaining the stability of the network through the "patch-bridge" mechanism only is not enough. By comparing the average pore diameter (225 μm) of the cellulose foam after curing without gelatin with the average bubble diameter (113 μm) of the wet foam under the same conditions, it is found that the bubbles not only become thicker but are also subjected to other strong destructive forces during the curing process. For example, in the process of freezing to generate the ice crystals, the force generated by the expansion of the growth volume of the ice crystals pushes the cellulose [50]. This results in the destruction of the cellulose network that is inherently unstable and has a larger pore size. In contrast, based on the temperature-controlled cross-linking properties of the gelatin, gelation occurs rapidly at low temperature after the addition of gelatin [51]. Moreover, the liquid bridges connected between the cellulose are converted into solid bridges, which allows the foam structure to be fixed before it can be coarsened. Not only is the original network structure of the wet foam maintained, but also the gelatin elastic aggregates tend to be connected through the forces of the polar group interactions such as hydrogen bonds and so on during the aging process. At the same time, the gelatin elastic aggregates are pulled into the overlapping distance that exists between the cellulose [52]. Therefore, it reduces the pore size of the foam. When the gelatin concentration is further increased to 3.2%, the liquid bridge endows too much force to the cellulose wet foam. The cross-linking between the gelatin and the gelatin molecules hinder the interaction between the gelatin and the cellulose. This results in the stratification of the foam, the destruction of network structure, and the increase in the foam pore size. The results show that the second liquid concentration has a significant effect on the pore size distribution of the cellulose-based foam. When the volume fraction of gelatin is 2.4%, the pore size of the cellulose-based foam reaches the minimum value.

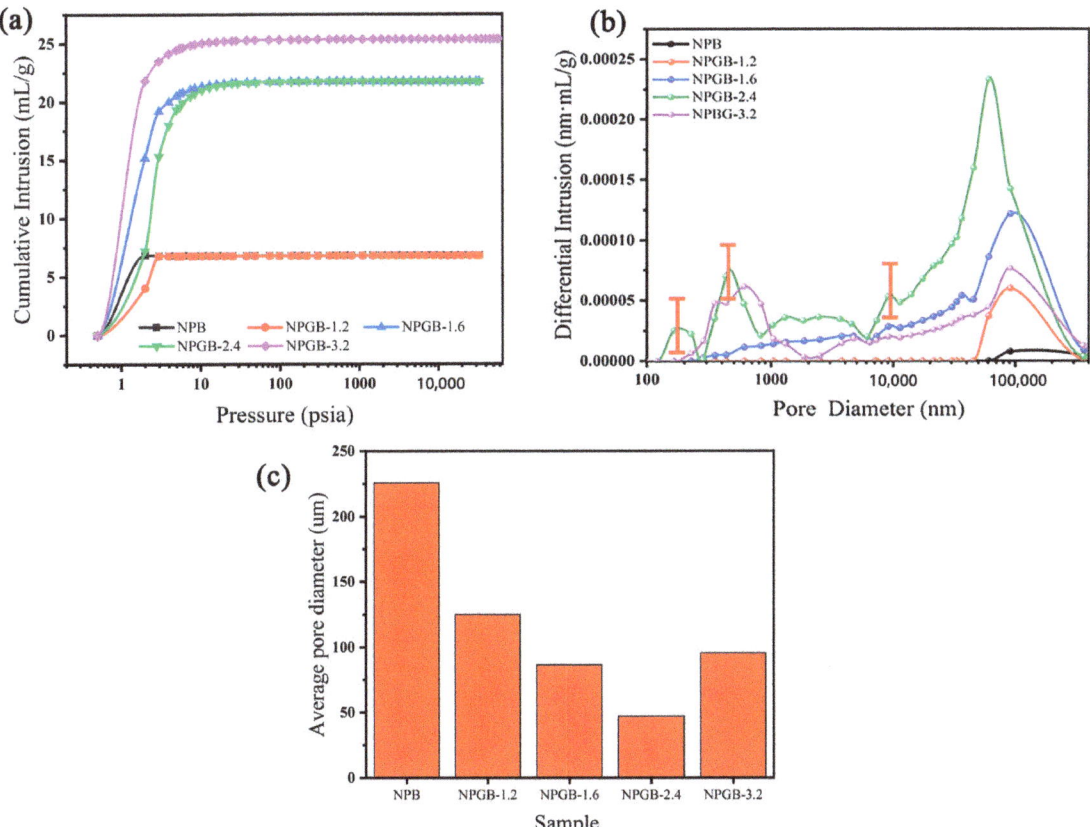

Figure 2. (**a**) Mercury injection curve, (**b**) pore size distribution, and (**c**) average pore size of cellulose-based foam with different gelatin volume fractions.

3.3. Crystal Structure of the Cellulose-Based Foam

The crystal structure of the cellulose-based foam is studied and analyzed by an X-ray diffractometer. The addition of the amorphous gelatin can reduce the ordered structure of the foam to a certain extent. As shown in Figure 3a, the main diffraction peaks appear at $2\theta = 16.5$ and $22.5°$, corresponding to the crystal planes (110) and (200), respectively. This figure conforms to the type I structure of typical cellulose [53]. For NPGB, the characteristic peaks are similar to NPB, indicating that the crystal structure of cellulose did not change after the addition of gelatin. When the volume fraction of gelatin is increased from 0% to 2.4%, the crystallinity of the foam decreases from 76.81% to 28.20%, and when the volume fraction of gelatin is further increased, the crystallinity (CrI) tends to increase too, as shown in Figure 3b. To analyze the reasons behind this pattern, the addition of amorphous gelatin may not only reduce the crystallinity of the foam but also contribute to the formation of the three-dimensional network structure of the cellulose-based foam [7]. With the addition of gelatin, the orientation of the cellulose that always tends to spread in a plane is partially broken, and the circular cell structure is gradually presented in the longitudinal direction. The foam converts from ordered to disordered, so the overall crystallinity decreases. When the volume fraction of the gelatin reaches 2.4%, the three-dimensional network structure of the cellulose-based foam is completely formed. At this stage, the disorder of the foam reaches its maximum and the crystallinity of the foam reaches its minimum. When the concentration of the gelatin further increases, the foam undergoes delamination in the longitudinal direction and the three-dimensional structure is damaged. Thus, the crystallinity of the foam increases and its disorder decreases.

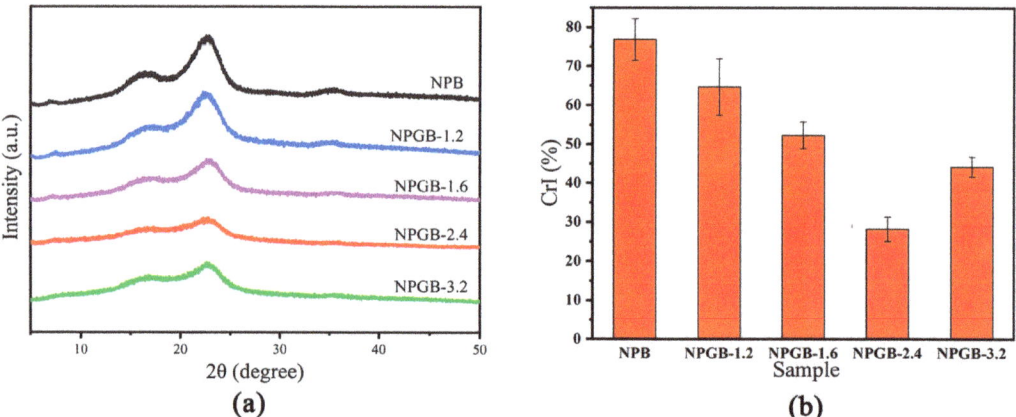

Figure 3. (a) XRD spectra and (b) crystallinity index (CrI) of cellulose-based foam with different gelatin volume fractions.

3.4. Mechanical Properties of the Cellulose-Based Foam

The stress–strain curve can reflect the change in the internal structure of the foam under the action of a force and the foam's response to it [54]. From the stress–strain curve shown in Figure 4, it can be seen that there are two distinct response regions. However, Chen and Kobayashi et al. [55] suggest the presence of three areas, in which the elastic deformation of the pore wall and the compression of the macropores mainly occur in the linear elastic area, the plastic yield of the pore wall occurs in the platform area, and the densification of the foam porous structure often occurs in the densification area. According to the literature [56], this is related to the processing route of the materials. For example, the lightweight foam which Cervin et al. [8] prepared through Pickering foam does not have a platform area, and when the strain reaches 60%, it directly switches from the linear elastic area to the dense area.

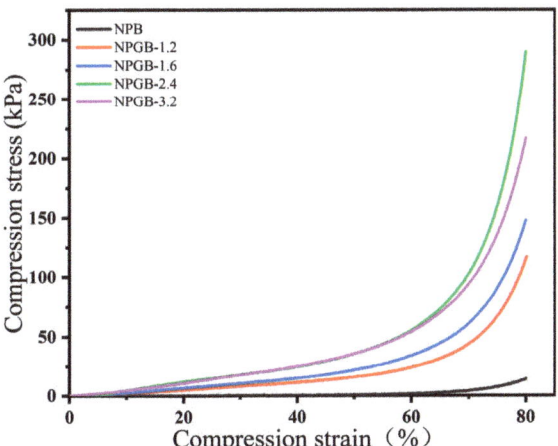

Figure 4. Stress–strain curve of cellulose-based foam with the different gelatin volume fractions.

In this study, the same behavior was observed in the above-mentioned report. Moreover, before the deformation reaches 60%, the stress increases linearly with the strain in the linear elastic region. Then, the stress extends to the densified area, and no platform appears. From the curves in Figure 4, it can be seen that the stress of the solid foam increases at first, and then decreases under the same deformation conditions with the increase in gelatin volume fraction in the linear elastic region. The stress reaches the maximum value when the gelatin volume fraction is 2.4%. The compressive stress values at 60% strain for NPB, NPGB-1.2, NPGB-1.6, NPGB-2.4, and NPGB-3.2 are 1.35546 kPa, 24.13049 kPa, 34.23665 kPa, 55.74601 kPa, and 54.50891 kPa, respectively. Furthermore, at 80% strain, the respective stress values reach 14.03921 kPa, 116.98721 kPa, 148.10305 kPa, 289.73374 kPa, and 216.98205 kPa. These are higher than the previously reported values for cellulose foam/aerogel shown in Table 2. When the volume fraction of gelatin increases up to 3.2%, the stress decreases slightly. This may be caused by the cross-linking of the excessive gelatin, which results in a vertical stratification. This destroys the three-dimensional network of cellulose-based foam, leading to stress concentration and decreased mechanical strength.

Table 2. Summary of the properties of the physical parameters of the cellulose-based foam materials.

Material	Density (g/cm^3)	Porosity (%)	Mechanical Strength (kPa)	References
Novel cellulose foam	0.096–0.0175	≥98%	55.746 (60%)	
NFC/MFC foam	0.010–0.060	90.0	13.78	[29]
NFC aerogels	<0.030	99.7	13.78	[57]
Silanized NFC sponge	0.017	99.0	27.70 (50%)	[58]
NFC foam	0.010	99.4	12.00 (50%)	[59]
Lignin/cellulose	0.010	80.0–90.0	200.00	[60]
Cellulose foam	0.020–0.065	80.0–90.0	10.00–90.00	[61]
Nano cellulose foam	0.011	97.1–99.4	60.00 (80%)	[62]
Cellulose scaffold	0.006–0.176	99.7	271.00 (70–80%)	[63]

3.5. Adsorption and Cycle Performance of the Cellulose-Based Foam

The driving force of the superhydrophobic/superoleophilic cellulose foam in absorbing oil is derived from the capillary force of the superhydrophobic hierarchical structures [64]. As shown in Figure 5a, the highly porous cellulose-based foams have an absorption capacity of 20 to 60 g/g for the various organic solvents or oils. They mainly absorb the liquid through the capillary action and physical sealing in the narrow spaces of the pores.

Due to the loose structure and the difficulty in forming the gelatin cellulose-based foam (NPB), this chapter only investigates the adsorption capacity of gelatin cellulose-based foam (NPGB). By comparing the absorption capacities of NPGB-1.2, NPGB-1.6, NPGB-2.4, and NPGB-3.2 for different solvents, it is found that when the volume fraction of gelatin is 1.2%, 1.6%, or 2.4%, there is no significant difference in the absorption capacity of the foam, and the maximum absorption capacities are 52.040 g/g, 55.625 g/g, and 57.061 g/g, respectively. In contrast, when the volume fraction of gelatin reaches 3.2%, the absorption capacity of the foam reduced significantly, and the maximum absorption capacity is 31.606 g/g, which may be due to the damage of the three-dimensional structure of the foam.

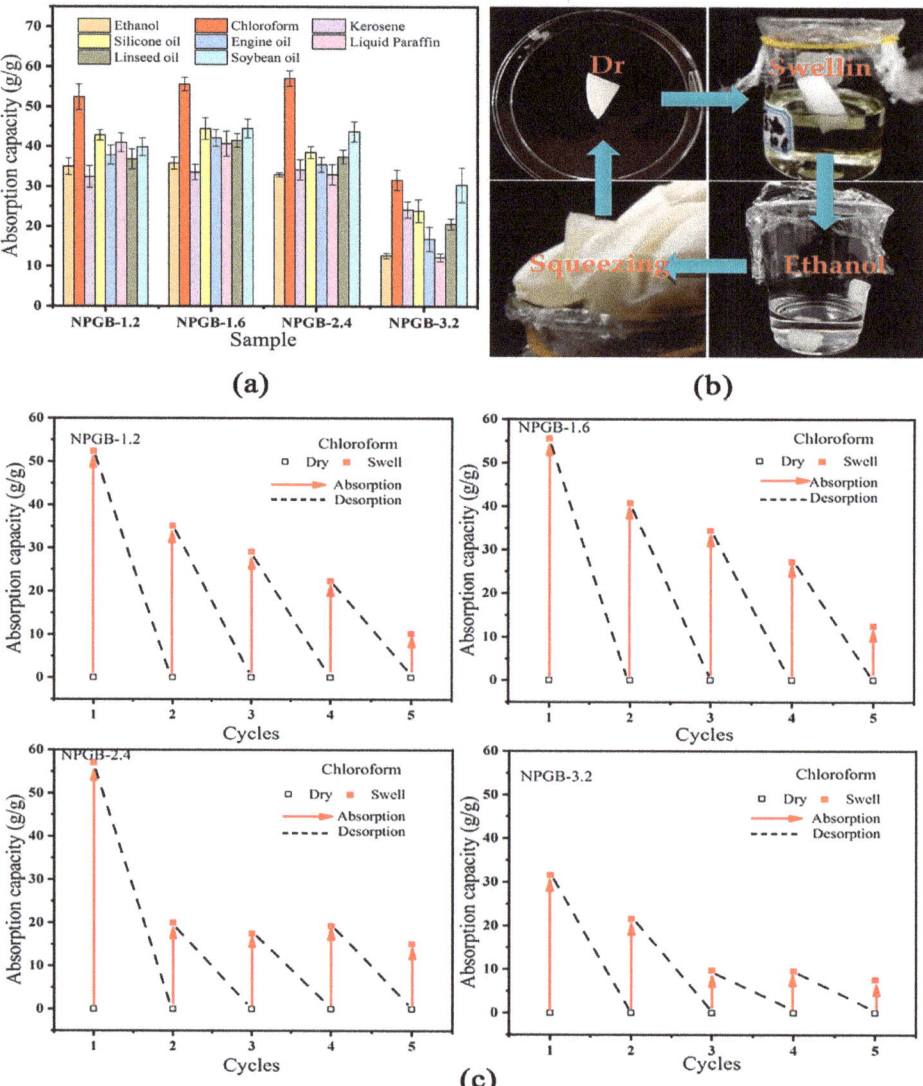

Figure 5. (**a**) Absorption capacity of foam for different organic solvents/oils, (**b**) the flowchart of the cycle for adsorption–desorption, and (**c**) the adsorption capacity of the foam for different organic solvents/oils at different gelatin volume fractions.

The traditional extrusion–drying–absorption cycle method can be used to test the cycle performance of the cellulose-based foam. The specific process is shown in Figure 5b. A certain volume of the foam is placed in different organic solvents/oils. After adsorption, the adsorbed solvent is extruded by simple extrusion and the foam is soaked in ethanol overnight for desorption. The foam is then transferred to a vacuum drying oven and dried at 60 °C for 10 h. This process is called a cycle. The cellulose-based foam is a kind of porous adsorption material which plays an important role in the recycling of materials and the adsorption of substances.

Figure 5c shows the absorption efficiency of a cellulose-based foam containing different gelatin volume fractions at different cycles, using chloroform as an example. It can be seen that the foam undergoes one adsorption, and the absorption amount decreases by 20–65% during the next adsorption. After five adsorption cycles, the adsorption capacity decreased by 73%. This may be analyzed by two factors. On the first hand, in the first adsorption process of the foam, the organic solvents and the oil enter the foam holes and stay there causing a blockage, so they become difficult to completely desorb. Therefore, no additional liquid storage space remains for the next adsorption. On the other hand, when desorption occurs after the last adsorption is complete, although a large amount of the solvent adsorbed in the foam can be discharged by the extrusion method, the structure of the foam is severely damaged. Thus, it results in a decrease in the adsorption capacity.

By comparing the circulating capacities of the foams with different gelatin concentrations, we found that the adsorption capacity of the foam cycle when the volume fraction of the gelatin is 2.4% or 3.2% is worse than that when the volume fraction is 1.2% or 1.6%. The reason is analyzed according to the surface morphology and the size distribution of the foam. The addition of the gelatin can make the cell wall of the cellulose-based foam thicker and the cell smaller. Where the mechanical strength and the water retention capacity of the foam are improved. After one round of adsorption, it is more difficult to desorb the oil droplets from the foam, and the recovery capacity of the foam decreases. The 3.2% gelatin volume fraction is the worst among the tested volume fractions in terms of liquid absorption and circulation. The results show that the gelatin concentration has an important effect on the adsorption and the recyclability of the cellulose-based foam. As the gelatin concentration increases, the adsorption cycle capacity decreases.

4. Conclusions

In this study, BF, NFC, TAD, and GAC are used as the raw materials, and SDS is used as a green solvent to prepare the cellulose-based foams with different gelatin concentrations via capillary foam technology. The results show that the structure of cellulose-based foam after adding gelatin is obviously compact, the crystallinity is decreased, the disorder is increased, and the mechanical properties are improved, but the recycling ability is decreased. Furthermore, the results show that the mechanical properties of the cellulose-based foam are improved, but its circulation capacity is decreased. The mechanical properties of the cellulose-based foam improve with the increase in the gelatin volume fraction. However, excessive addition of gelatin causes excessive crosslinking of foam, which destroys the three-dimensional network of the cellulose-based foam, leading to stress concentration and decrease in the mechanical strength, and the experimental results show that when the volume fraction of gelatin is 2.4%, the stress is 55.746 kPa. At this time, the mechanical properties of the cellulose-based foam are the best, and the adsorption capacity of the foam also reaches the highest 57.061 g/g. In addition, with the increase in gelatin concentration, the circulation capacity of the cellulose-based foam decreases, and the number of adsorption cycles is limited, among which, the adsorption capacity decreased by 73% after five adsorption cycles. In addition, with the increase in gelatin concentration, the adsorption and circulation capacity of gelatin decreases. Therefore, the concentration of the secondary liquid (gelatin) has a great influence on the mechanical properties as well as the adsorption and circulation capacity of the cellulose-based foam.

Author Contributions: Conceptualization, Y.Z. and W.Y.; methodology, Y.L.; software, W.Y.; validation, Y.Z., W.Y. and C.Q.; formal analysis, Y.G.; investigation, Y.Q.; resources, Y.L.; data curation, Y.G.; writing—original draft preparation, C.Q.; writing—review and editing, Y.Z.; visualization, Y.Z.; supervision, Y.Q.; project administration, Y.L.; funding acquisition, W.Y. All authors have read and agreed to the published version of the manuscript.

Funding: National Natural Science Foundation of China (Grant No. 22068004).

Institutional Review Board Statement: Not applicable.

Data Availability Statement: No new data were created or analyzed in this study. Data sharing is not applicable to this article.

Acknowledgments: The authors want to thank National Natural Science Foundation of China (Grant No. 22068004) for the financial support to this research.

Conflicts of Interest: The authors declare no conflict of interest.

References

1. Gokana, S.; Rani, M.; Pathania, D.; Abhimanyu; Umapathi, R.; Rustagi, S.; Huh, Y.S.; Gupta, V.K.; Kaushik, A.; Chaudhary, V. Agro-waste to sustainable energy: A green strategy of converting agricultural waste to nano-enabled energy applications. *Sci. Total Environ.* **2023**, *875*, 162667.
2. Al-Qararah, A.M.; Ekman, A.; Hjelt, T.; Kiiskinen, H.; Timonen, J.; Ketoja, J.A. Porous structure of fibre networks formed by a foaming process: A comparative study of different characterization techniques. *J. Microsc.* **2016**, *264*, 88–101. [CrossRef] [PubMed]
3. Bhandari, J.; Mishra, H.; Mishra, P.K.; Wimmer, R.; Ahmad, F.J.; Talegaonkar, S. Cellulose nanofiber aerogel as a promising biomaterial for customized oral drug delivery. *Int. J. Nanomed.* **2017**, *12*, 2021–2031. [CrossRef] [PubMed]
4. Cervin, N.T.; Aulin, C.; Larsson, P.T.; Wågberg, L. Ultra porous nanocellulose aerogels as separation medium for mixtures of oil/water liquids. *Cellulose* **2012**, *19*, 401–410. [CrossRef]
5. Chen, B.; Zheng, Q.F.; Zhu, J.L.; Li, J.; Cai, Z.; Chen, L.; Gong, S. Mechanically strong fully biobased anisotropic cellulose aerogels. *Rsc. Adv.* **2016**, *6*, 96518–96526. [CrossRef]
6. de Oliveira, J.P.; Bruni, G.P.; El Halal, S.L.M.; Bertoldi, F.C.; Dias, A.R.G.; Zavareze, E.D.R. Cellulose nanocrystals from rice and oat husks and their application in aerogels for food packaging. *Int. J. Biol. Macromol.* **2019**, *124*, 175–184. [CrossRef]
7. Fan, X.; Li, Y.; Li, X.; Wu, Y.; Tang, K.; Liu, J.; Zheng, X.; Wan, G. Injectable antibacterial cellulose nanofiber/chitosan aerogel with rapid shape recovery for noncompressible hemorrhage. *Int. J. Biol. Macromol.* **2020**, *154*, 1185–1193. [CrossRef]
8. Cervin, N.T.; Johansson, E.; Larsson, P.A.; Wågberg, L. Strong, Water-Durable, and Wet-Resilient Cellulose Nanofibril-Stabilized Foams from Oven Drying. *Acs Appl. Mater. Interfaces* **2016**, *8*, 11682–11689. [CrossRef]
9. Arabkhani, P.; Asfaram, A. Development of a novel three-dimensional magnetic polymer aerogel as an efficient adsorbent for malachite green removal. *J. Hazard. Mater.* **2020**, *384*, 121394. [CrossRef]
10. Ma, S.; Zhang, M.; Nie, J.; Tan, J.; Song, S.; Luo, Y. Lightweight and porous cellulose-based foams with high loadings of zeolitic imidazolate frameworks-8 for adsorption applications. *Carbohydr. Polym.* **2019**, *208*, 328–335. [CrossRef]
11. Zhou, G.; Luo, J.; Liu, C.; Chu, L.; Ma, J.; Tang, Y.; Zeng, Z.; Luo, S. A highly efficient polyampholyte hydrogel sorbent based fixed-bed process for heavy metal removal in actual industrial effluent. *Water Res.* **2016**, *89*, 151–160. [CrossRef]
12. Mo, L.; Pang, H.; Lu, Y.; Li, Z.; Kang, H.; Wang, M.; Zhang, S.; Li, J. Wood-inspired nanocellulose aerogel adsorbents with excellent selective pollutants capture. superfast adsorption, and easy regeneration. *J. Hazard. Mater.* **2021**, *415*, 125612. [CrossRef] [PubMed]
13. Xu, T. Preparation and Properties of Sugarcane Cellulose Filament Porous Materials. Master's Thesis, Guangxi University, Nanning, China, 2018.
14. Klemm, D.; Heublein, B.; Fink, H.; Bohn, A. Cellulose: Fascinating biopolymer and sustainable raw material. *Angew. Chem.* **2005**, *44*, 3358–3393. [CrossRef]
15. Bhat, A.H.; Khan, I.; Usmani, M.A.; Umapathi, R.; Al-Kindy, S.M.Z. Cellulose an ageless renewable green nanomaterial for medical applications: An overview of ionic liquids in extraction, separation and dissolution of cellulose. *Int. J. Biol. Macromol.* **2019**, *129*, 750–777. [CrossRef] [PubMed]
16. Taha, M.; Silva, F.A.; Quental, M.V.; Ventura, S.P.M.; Freire, M.G.; Coutinho, J.A.P. Good's buffers as a basis for developing self-buffering and biocompatible ionic liquids for biological research. *Green Chem.* **2014**, *16*, 3149–3159. [CrossRef] [PubMed]
17. Khan, I.; Kurnia, K.A.; Sintra, T.E.; Saraiva, J.A.; Pinho, S.P.; Coutinho, J.A.P. Assessing the activity coefficients of water in cholinium-based ionic liquids: Experimental measurements and COSMO-RS modeling. *Fluid Phase Equilib.* **2014**, *361*, 16–22. [CrossRef]
18. Froschauer, C.; Hummel, M.; Iakovlev, M.; Roselli, A.; Schottenberger, H.; Sixta, H. Separation of hemicellulose and cellulose from wood pulp by means of ionic liquid cosolvent systems. *Biomacromolecules* **2013**, *14*, 1741–1750. [CrossRef] [PubMed]
19. Zhuo, K.; Du, Q.; Bai, G.; Wang, C.; Chen, Y.; Wang, J. Hydrolysis of cellulose catalyzed by novel acidic ionic liquids. *Carbohydr. Polym.* **2015**, *115*, 49–53. [CrossRef]

20. Ren, Q. Research on the Solubility Property of Cellulose in Ionic Liquid. Master's Thesis, University of Aeronautics & Astronautics, Beijing, China, 2003.
21. Feng, L.; Chen, Z.-L. Research progress on dissolution and functional modification of cellulose in ionic liquids. *J. Mol. Liq.* **2008**, *142*, 1–5. [CrossRef]
22. Hou, Q.P. Basic Research on Application of Foam Forming. Master's Thesis, South China University of Technology, Guangzhou, China, 2018.
23. Xiang, W.; Preisig, N.; Ketola, A.; Tardy, B.L.; Bai, L.; Ketoja, J.A.; Stubenrauch, C.; Rojas, O.J. How Cellulose Nanofibrils Affect Bulk, Surface, and Foam Properties of Anionic Surfactant Solutions. *Biomacromolecules* **2019**, *20*, 4361–4369. [CrossRef]
24. Chen, X.; Chen, Y.; Zou, L.; Zhang, X.; Dong, Y.; Tang, J.; McClements, D.J.; Liu, W. Plant-Based Nanoparticles Prepared from Proteins and Phospholipids Consisting of a Core-Multilayer-Shell Structure: Fabrication, Stability, and Foamability. *J. Agric. Food Chem.* **2019**, *67*, 6574–6584. [CrossRef] [PubMed]
25. Nechita, P.; Nastac, S.M. Overview on Foam Forming Cellulose Materials for Cushioning Packaging Applications. *Polymers* **2022**, *14*, 1963. [CrossRef]
26. Du, X.; Zhao, L.; Xiao, H.; Liang, F.; Chen, H.; Wang, X.; Wang, J.; Qu, W.; Lei, Z. Stability and shear thixotropy of multilayered film foam. *Colloid Polym. Sci.* **2014**, *292*, 2745–2751. [CrossRef]
27. Jiang, Q.; Bismarck, A. A perspective: Is viscosity the key to open the next door for foam templating. *React. Funct. Polym.* **2021**, *162*, 104877. [CrossRef]
28. Chen, S.Y.; Liu, H.J.; Yang, J.J. Bulk foam stability and rheological behavior of aqueous foams prepared by clay particles and alpha olefin sulfonate. *J. Mol. Liq.* **2019**, *291*, 111250. [CrossRef]
29. Liu, Y.; Lu, P.; Xiao, H.; Heydarifard, S.; Wang, S. Novel aqueous spongy foams made of three-dimensionally dispersed wood-fiber: Entrapment and stabilization with NFC/MFC within capillary foams. *Cellulose* **2017**, *24*, 241–251. [CrossRef]
30. Swerin, A. Rheological properties of cellulosic fibre suspensions flocculated by cationic polyacrylamides. *Colloids Surf. A-Physicochem. Eng. Asp.* **1998**, *133*, 279–294. [CrossRef]
31. Al-Qararah, A.M.; Hjelt, T.; Koponen, A.; Harlin, A.; Ketoja, J.A. Bubble size and air content of wet fibre foams in axial mixing with macro-instabilities. *Colloids Surf. A Physicochem. Eng. Asp.* **2013**, *436*, 1130–1139. [CrossRef]
32. Lappalainen, L.; Lehmonen, J. Determinations of bubble size distribution of foam-fibre mixture using circular hough transform. *Nord. Pulp Pap. Res. J.* **2012**, *27*, 930–939. [CrossRef]
33. Koponen, A.I.; Oleg, T.; Ari, J.H.; Kiiskinen, H. Drainage of high-consistency fiber-laden aqueous foams. *Cellulose* **2020**, *27*, 9637–9652. [CrossRef]
34. Li, H. Study on the Construction and Flow Behavior of Nano-Cellulose Reinforced Foam System. Master's Thesis, Southwest University of Petroleum, Chengdu, China, 2018.
35. Szopa, D.; Mielczarek, M.; Skrzypczak, D.; Izydorczyk, G.; Mikula, K.; Chojnacka, K.; Witek-Krowiak, A. Encapsulation efficiency and survival of plant growth-promoting microorganisms in an alginate-based matrix-A systematic review and protocol for a practical approach. *Ind. Crops Prod.* **2022**, *181*, 114846. [CrossRef]
36. Huprikar, S.; Usgaonkar, S.; Lele, A.K.; Orpe, A.V. Microstructure and yielding of capillary force induced gel. *Rheol. Acta* **2020**, *59*, 291–306. [CrossRef]
37. Dittmann, J.; Koos, E.; Willenbacher, N. Ceramic Capillary Suspensions: Novel Processing Route for Macroporous Ceramic Materials. *J. Am. Ceram. Soc.* **2013**, *96*, 391–397. [CrossRef]
38. Ishigami, T.; Tokishige, C.; Fukasawa, T.; Fukui, K.; Kihara, S. Semiphenomenological model to predict hardening of solid-liquid-liquid systems by liquid bridges. *Granul. Matter* **2019**, *21*, 103. [CrossRef]
39. Xue, R.; Zhao, H.; An, Z.W.; Wu, W.; Jiang, Y.; Li, P.; Huang, C.X.; Shi, D.; Li, R.; Hu, G.H.; et al. Self-healable, solvent response cellulose nanocrystal/waterborne polyurethane nanocomposites with encryption capability. *ACS Nano.* **2023**, *17*, 5653–5662. [CrossRef]
40. An, Z.W.; Ye, K.; Xue, R.; Zhao, H.; Liu, Y.; Li, P.; Chen, Z.; Huang, C.; Hu, G.H. Recent advances in self-healing polyurethane based on dynamic covalent bonds combined with other self-healing methods. *Nanoscale* **2023**. [CrossRef]
41. Domenech, T.; Velankar, S.S. On the rheology of pendular gels and morphological developments in paste-like ternary systems based on capillary attraction. *Soft Matter.* **2015**, *11*, 1500–1516. [CrossRef]
42. Cervin, N.T.; Andersson, L.; Ng, J.B.; Olin, P.; Bergström, L.; Wågberg, L. Lightweight and Strong Cellulose Materials Made from Aqueous Foams Stabilized by Nanofibrillated Cellulose. *Biomacromolecules* **2013**, *14*, 503–511. [CrossRef]
43. Ferreira, E.S.; Cranston, E.D.; Rezende, C.A. Naturally Hydrophobic Foams from Lignocellulosic Fibers Prepared by Oven-Drying. *Acs Sustain. Chem. Eng.* **2020**, *8*, 8267–8278. [CrossRef]
44. Wang, P.; Aliheidari, N.; Zhang, X.; Ameli, A. Strong ultralight foams based on nanocrystalline cellulose for high-performance insulation. *Carbohydr. Polym.* **2019**, *218*, 103–111. [CrossRef]
45. Yao, M.Z.; Liu, Y.; Qing, C.N.; Hui, Z.; Huang, Q.X. Effect of hemicellulose content on structure and properties of bagasse fiber-based membrane. *J. Packag. Eng.* **2020**, *41*, 60–66.
46. Zhu, H.G. Preparation of Functionalized Micro-Nano Composite Based on Graphene and Its Application in Water Purification. Ph.D. Thesis, Suzhou University, Suzhou, China, 2018.
47. Wang, L.W. High Strength Microfibrillated Cellulose/Gelatin Composite Hydrogel with Controllable Network Structure. Ph.D. Thesis, Zhengzhou University, Zhengzhou, China, 2018.

48. Guo, L.; Chen, Z.; Lyu, S.; Fu, F.; Wang, S. Highly flexible cross-linked cellulose nanofibril sponge-like aerogels with improved mechanical property and enhanced flame retardancy. *Carbohydr. Polym.* **2018**, *179*, 333–340. [CrossRef] [PubMed]
49. Li, J.; Lu, Z.; Xie, F.; Huang, J.; Ning, D.; Zhang, M. Highly compressible, heat-insulating and self-extinguishing cellulose nanofiber/aramid nanofiber nanocomposite foams. *Carbohydr. Polym.* **2021**, *261*, 117837. [CrossRef] [PubMed]
50. Liu, Q. Study on Pore Structure Regulation Technology of Nano-Cellulose-Based Foam Material. Ph.D. Thesis, Shanxi University of Science and Technology, Xianyang, China, 2017.
51. Ji, Z.L. Studies on the Compatibility and Phase Behavior of Gelatin/Hydroxypropyl Methylcellulose Blends. Ph.D. Thesis, South China University of Technology, Guangzhou, China, 2020.
52. Wang, J.R. Preparation and Properties of Gelatin-Based High-Strength Hydrogel. Ph.D. Thesis, Zhengzhou University, Zhengzhou, China, 2020.
53. Lu, Y.X. Preparation and Performance Study of Friction Nano-Generator Based on Cellulose Nanofiber Composites. Ph.D. Thesis, Guangxi University, Nanning, China, 2020.
54. Lavoine, N.; Bergstrom, L. Nanocellulose-based foams and aerogels: Processing, properties, and applications. *J. Mater. Chem.* **2017**, *5*, 16105–16117. [CrossRef]
55. Yang, X.; Cranston, E.D. Chemically Cross-Linked Cellulose Nanocrystal Aerogels with Shape Recovery and Superabsorbent Properties. *Chem. Mater.* **2014**, *26*, 6016–6025. [CrossRef]
56. Qin, C.; Yao, M.; Liu, Y.; Yang, Y.; Zong, Y.; Zhao, H. MFC/NFC-Based Foam/Aerogel for Production of Porous Materials: Preparation, Properties and Applications. *Materials* **2020**, *13*, 5568. [CrossRef]
57. Aulin, C.; Netrval, J.; Wagberg, L.; Lindström, T. Aerogels from nanofibrillated cellulose with tunable oleophobicity. *Soft Matter.* **2010**, *6*, 3298–3305. [CrossRef]
58. Zheng, Z.; Sèbe, G.; Rentsch, D.; Zimmermann, T.; Tingaut, P. Ultralightweight and Flexible Silylated Nanocellulose Sponges for the Selective Removal of Oil from Water. *Chem. Mater.* **2014**, *26*, 2659–2668. [CrossRef]
59. Antonini, C.; Wu, T.; Zimmermann, T.; Kherbeche, A.; Thoraval, M.J.; Nyström, G.; Geiger, T. Ultra-Porous Nanocellulose Foams: A Facile and Scalable Fabrication Approach. *Nanomaterials* **2019**, *9*, 1142. [CrossRef]
60. Kobayashi, Y.; Saito, T.; Isogai, A. Aerogels with 3D ordered nanofiber skeletons of liquid-crystalline nanocellulose derivatives as tough and transparent insulators. *Angew. Chem. Int. Ed.* **2015**, *53*, 10253. [CrossRef]
61. Poehler, T.; Jetsu, P.; Isomoisio, H. Benchmarking new wood fibre-based sound absorbing material made with a foam-forming technique. *Build. Acoust.* **2016**, *23*, 131–143. [CrossRef]
62. Martoia, F.; Cochereau, T.; Dumont, P.J.J.; Orgéas, L.; Terrien, M.; Belgacem, M.N. Cellulose nanofibril foams: Links between ice-templating conditions, microstructures and mechanical properties. *Mater. Des.* **2016**, *104*, 376–391. [CrossRef]
63. Liu, J.; Cheng, F.; Grénman, H.; Spoljaric, S.; Seppälä, J.; Eriksson, J.; Willför, S.; Xu, C. Development of nanocellulose scaffolds with tunable structures to support 3D cell culture. *Carbohydr. Polym. Sci. Technol. Asp. Ind. Important Polysacch.* **2016**, *145*, 259–271. [CrossRef] [PubMed]
64. Liu, C.-H.; Shang, J.-P.; Su, X.; Zhao, S.; Peng, Y.; Li, Y.-B. Fabrication of Superhydrophobic/Superoleophilic Bamboo Cellulose Foam for Oil/Water Separation. *Polymers* **2022**, *14*, 5162. [CrossRef] [PubMed]

Disclaimer/Publisher's Note: The statements, opinions and data contained in all publications are solely those of the individual author(s) and contributor(s) and not of MDPI and/or the editor(s). MDPI and/or the editor(s) disclaim responsibility for any injury to people or property resulting from any ideas, methods, instructions or products referred to in the content.

Article

Preparation of Bio-Based Foams with a Uniform Pore Structure by Nanocellulose/Nisin/Waterborne-Polyurethane-Stabilized Pickering Emulsion

Yiqi Chen, Yujie Duan, Han Zhao, Kelan Liu, Yiqing Liu, Min Wu and Peng Lu *

College of Light Industry and Food Engineering, Guangxi University, Nanning 530004, China
* Correspondence: author: lupeng-1984@163.com

Abstract: Bio-based porous materials can reduce energy consumption and environmental impact, and they have a possible application as packaging materials. In this study, a bio-based porous foam was prepared by using a Pickering emulsion as a template. Nisin and waterborne polyurethane (WPU) were used for physical modification of 2,2,6,6-tetramethyl piperidine-1-oxyl-oxidized cellulose nanocrystals (TOCNC). The obtained composite particles were applied as stabilizers for acrylated epoxidized soybean oil (AESO) Pickering emulsion. The stability of the emulsion was characterized by determination of the rheological properties and microscopic morphology of the emulsion. The emulsion stabilized by composite particles showed better stability compared to case when TOCNC were used. The porous foam was obtained by heating a composite-particles-stabilized Pickering emulsion at 90 °C for 2 h. SEM (scanning electron microscopy) images showed that the prepared foam had uniformly distributed pores. In addition, the thermal conductivity of the foam was 0.33 W/m·k, which was a significant decrease compared to the 3.92 W/m·k of the TOCNC foam. The introduction of nisin and WPU can reduce the thermal conductivity of the foam, and the physically modified, TOCNC-stabilized Pickering emulsion provides an effective means to preparing bio-based porous materials.

Keywords: TOCNC/Nisin/WPU; Pickering emulsion; biomaterials; porous foam; thermal insulation

Citation: Chen, Y.; Duan, Y.; Zhao, H.; Liu, K.; Liu, Y.; Wu, M.; Lu, P. Preparation of Bio-Based Foams with a Uniform Pore Structure by Nanocellulose/Nisin/Waterborne-Polyurethane-Stabilized Pickering Emulsion. *Polymers* **2022**, *14*, 5159. https://doi.org/10.3390/polym14235159

Academic Editor: Pablo Marcelo Stefani

Received: 18 October 2022
Accepted: 23 November 2022
Published: 27 November 2022

Publisher's Note: MDPI stays neutral with regard to jurisdictional claims in published maps and institutional affiliations.

Copyright: © 2022 by the authors. Licensee MDPI, Basel, Switzerland. This article is an open access article distributed under the terms and conditions of the Creative Commons Attribution (CC BY) license (https://creativecommons.org/licenses/by/4.0/).

1. Introduction

Porous foams are a class of materials with high porosities and large high specific surface areas, which are widely used in thermal insulation, energy storage, etc. At present, the production of most foams applied in the commercial field highly depends on petroleum resources, which limits their application. As fossil and petroleum resources become increasingly depleted, it is becoming a trend to use renewable resources and environmentally friendly production strategies to develop bio-based foams to replace traditional foam materials [1]. Various biomaterials have been reported for the preparation of bio-based foams. Hassan et al. [2] fabricated a biodegradable starch/cellulose composite foam with starch-containing cellulose fibers as a reinforcing agent and citric acid as a cross-linking agent. The composite foams can be applied as a biodegradable replacement for polystyrene foam. Qiu et al. [3] used AESO chemically grafted with flame retardants to produce a bio-based foam with desirable mechanical properties and flame retardancy. Luo et al. [4] reported a biodegradable poly (3-hydroxybutyrate-co-3-hydroxyvalerate) (PHBV)-based electromagnetic shielding foam by supercritical CO_2, and constructed a green and economic avenue for the application of this PHBV foam in the field of electromagnetic shielding.

Emulsion is a dispersed system consisting of several immiscible liquids. Compared with traditional surfactant-stabilized emulsion, a Pickering emulsion generally has excellent stability due to nearly irreversible interfacial adsorption of stabilizers [5]. In addition, a Pickering emulsion also has environmentally friendly properties—e.g., a particle-stabilized emulsion can reduce the total amount of surfactant required. Based on the basic properties of a Pickering emulsion, it is an excellent template for preparing bio-based porous foam, and the pore structure of the foam can be adjusted by the stability of the emulsion and droplet size [6]. In a Pickering emulsion, particle stabilizers are particularly important for the stability of it and the properties of the prepared material. Some inorganic particles, such as nonmetallic oxide particles and magnetic particles, have been extensively studied in Pickering emulsions, but their common disadvantage is poor biocompatibility. In recent years, as a renewable material which has superior biocompatibility and biodegradability, nanocellulose, has attracted the interest of researchers.

Nanocellulose is a renewable and biodegradable nanomaterial that combines a large specific surface area, flexibility, low density and chemical inertness. Due to the presence of hydroxyl groups on the surface of nanocellulose, it can be physically modified or chemically modified with other polymers and nanomaterials by functional groups or grafting biomolecules [7]. In recent years, nanocellulose, which is carbon-neutral, non-toxic and sustainable, has attracted a lot of attention due to environmental concerns. At present, nanocellulose is widely applied in food packaging [8], thermal insulation materials [9,10], coatings [11], biomedicine [12], etc. Due to the surface properties, shape and inter-particle interactions of nanocellulose, it shows good self-assembly ability at the liquid interface and has been used to stabilize Pickering emulsions [13,14].The surface of nanocellulose has abundant OH groups, which gives it hydrophilicity overall, and the hydrophobic (200) β crystalline edges containing CH groups impart hydrophobicity [15]. The amphiphilicity of nanocellulose plays a crucial role in the stability of a Pickering emulsion. On the other hand, the stability of the emulsion generally affects the pore structure and properties of the prepared foam, including the thermal insulation and mechanical properties [7]. Zhang et al. [16] used aminated cellulose nanocrystals (CNC) to stabilize an oil-in-water (o/w) Pickering high internal phase emulsion (HIPE), and constructed a CNC aerogel with high porosity and low thermal conductivity by a simple Pickering emulsion template method. Capron et al. [17] reported a lightweight foam with porous structures which was obtained from freeze-drying the CNC stabilized o/w Pickering emulsion. Liu et al. [18] prepared a microwave absorbing foam by compounding the freeze-drying with the o/w Pickering emulsion gelation method, which emulsion was co-stabilized with CNF, carbon nanotubes (CNT) and Fe_3O_4 nanoparticles; and the prepared foams showed excellent thermal insulation properties compared to commercial polyvinyl alcohol and polyurethane foams. However, it still is a challenge to develop a strategy to prepare foams using nanocellulose-stabilized water-in-oil (w/o) Pickering emulsions.

In general, for a w/o Pickering emulsion, porous foams can be prepared through the polymerization and curing of the oil phase, and the function of foams can be adjusted by the choice of colloidal particles that are used as the stabilizer [19]. However, nanocellulose has high hydrophilicity and struggles to form a stable w/o Pickering emulsion by itself [20]. To improve the stability of nanocellulose in making Pickering emulsions, several attempts have been made, such as acetylation [21], quaternary surfactant adsorption [22] and organic-acid grafting [23].

Nisin is a cationic antimicrobial polypeptide with 34 amino acids which has been used as an antimicrobial agent in food. In recent years, nisin has mainly been combined with other polymeric matrices, such as polyethylene, nanocellulose, maize protein and starch, by physical and chemical methods [24]. Amino groups carried by nisin give it a positive charge. This allows it to be combined with other materials by electrostatic attraction, which gives it great potential for packaging material functionalization. Polyurethane (PU) is a copolymer containing repeated urethane groups composed of soft and hard chain segments. It has excellent mechanical properties, good compatibility and is easy to modify [25,26]. However, traditional PU could release a large amount of volatile organic compounds during its application. Therefore, environmentally friendly WPU is gaining more and more attention [27], and applying WPU to Pickering emulsions may be an interesting topic. Due to WPU's deformable particle structure and functional groups, it is a potential emulsifier for the preparation of Pickering emulsions, and WPU particles, as soft colloids, can impart deformable interfaces to a Pickering emulsion [28].

With growing concern around environmental issues, incorporating plant-derived materials into current polymer systems to create new bio-based polymers or polymer foams is an attractive research topic. Vegetable oils are considered ideal raw materials for the preparation of sustainable foams because they are abundant and readily available. AESO is a soybean oil derivative consisting of triglyceride oil, which can be polymerized into high-molecular-weight and highly cross-linked thermoset polymers. In recent years, AESO has been used as a resin to form "green" composites or foam materials [29,30].

Herein, we submit a method for the preparation of bio-based porous foams using a physically modified, TOCNC-stabilized AESO Pickering emulsion. In our study, cationic nisin was immobilized on the surface of anionic TOCNC by electrostatic interaction, providing a basis for enhancing the interface stability of TOCNC as a Pickering-foam stabilizer [31]. On the basis of this, WPU was introduced to synergistically enhance the stability of TOCNC in the preparation of Pickering emulsions. The stability of Pickering emulsions was evaluated by microscopic observations and rheological tests. A Pickering emulsion was used as a template to prepare porous foam by thermal curing of AESO. Furthermore, the pore structure and thermal insulation properties of the foam were studied.

2. Materials and Methods

2.1. Materials

Nisin ($\geq$1000 IU/mg) was purchased from Xinyinxiang Biological Engineering Co., Ltd. (Zhejiang, China). TOCNC were purchased from Zhejiang Jinjiahao Green Nanomaterials Co., Ltd. (Zhejiang, China). Hydrochloric acid and sodium hydroxide were purchased from Chengdu Kelong Chemical Reagent Factory (Sichuan, China). Dimethylolpropionic acid (DMPA), 1,4-butanediol (BDO), isophorone diisocyanate (IPDI), 4-hydroxyanisole, acetone, dibutyltin dilaurate, polybutylene glycols (PTMG), benzoyl peroxide (BPO), anhydrous acetone, triethylamine (TEA), epoxidized soybean oil (ESO), 4-methoxyphenol (MEHQ) and triphenylphosphine (TPP) were purchased from Aladdin Reagent Co., Ltd. (Shanghai, China). All chemical agents used in this research were analytical grade.

2.2. Preparation of WPU

WPU was synthesized in the laboratory using the following procedure. Firstly, a mixture of PTMG (30 g), acetone (60 g) and DMPA (2.68 g) were added into a four-necked flask equipped with a mechanical stirrer and thermometer. The reaction was carried out at 60 °C under argon for 20 min. Then IPDI (15.54 g) and dibutyltin dilaurate were added into the mixture, and the reaction was carried out at 80 °C for 4 h. After that, BDO (1.35 g) was added into flask, and the reaction was kept for 1 h. Subsequently, TEA (2.02 g) was added to neutralize the carboxylic groups in the prepolymer, and the temperature was reduced to 10 °C after 1 h. Finally, ultrapure water (103 g) was added into flask with a constant pressure funnel to obtain the WPU solution. The prepared WPU solution was stored at room temperature.

2.3. Preparation and Characterization of Particle Suspensions

Deionized water was added to the TOCNC aqueous dispersion and magnetically stirred for 30 min to prepare a TOCNC suspension (0.2 wt%). A TOCNC/nisin (TCN) suspension was prepared by adding a nisin solution (0.03 wt%) into a TOCNC suspension and magnetically stirring for 3 h. The TOCNC/nisin/WPU (TCNW) suspension was prepared by adding a WPU solution (0.01 wt%) into the TCN suspension and stirring for 1 h. All particle suspensions were prepared at pH = 7.

The transmission electron microscopy (TEM, HT7700, Hitachi, Japan) was used to observe the topographic characteristic of particles. The particle suspension was diluted to 0.005 wt% with deionized water; then a drop of particle suspension was deposited on a carbon-coated copper grid, followed by drying out overnight at 25 °C. The grid was observed using TEM at 100 kV.

The interfacial tension of the suspensions on the air–water interface was determined using a tensiometer (JK99F, Zhongchen Digital Technology Equipment Co., Ltd., Shanghai, China) at 25 °C and 60% relative humidity.

A high-speed blender (Unidrive 1000D, Ballrechten-Dottingen, Germany) was used to shear particle suspensions at 12,000 rpm for 2 min; then, the foaming heights were recorded at different times. The micromorphology of bubbles was studied by an optical microscope (Leica FSC, Leica Instruments Ltd., Weztlar, Germany).

2.4. Preparation and Characterization of Emulsions

The prepared particles were used as stabilizers for w/o Pickering emulsions. In our previous experiments, we used AESO Pickering emulsions as templates to prepare porous foams [19,32,33]. The effects of different particles and different oil–water ratios on the stability of emulsions were investigated by the characterization of centrifugal stability, rheological properties and viscosity of emulsions. In the experiments of using nanocellulose-stabilized AESO emulsions to the prepare foams, we found that the emulsions with water content of between 10–40% had better stability, so an oil–water ratio of 6:4 was chosen for preparing the Pickering emulsions in our study. Emulsions were prepared by mixing the oil phase with the particle suspension at a fixed oil-to-water ratio of 6:4 using a high-speed blender at 12,000 rpm for 3 min. Typically, a mixture of AESO, BPO (3 wt%), and anhydrous acetone was used as the oil phase. Among them, AESO was prepared according to the research method in [33].

The micromorphology of emulsions was studied using an optical microscope. The size distribution of emulsion droplets was determined by Nano Measurer 1.2 software.

The rheological properties of the emulsion were tested by a rotary rheometer (HAAKE MARS 40, Thermo Fisher, Karlsruhe, Germany). Oscillation amplitude frequency sweep with strain sweep from 0.1% to 100% was performed at 25 °C with an angle frequency of 1 rad/s to obtain the linear viscoelastic regions (LVR) of emulsions, and the limiting deformation value of the emulsion was given by the rheological system according to the variation in the storage modulus (G′) and the loss modulus (G″) within the LVR. The gap between two parallel plates (Φ = 35 mm) was set to 1 mm.

2.5. Preparation and Characterization of Foams

The foams were prepared by polymerization of AESO continuous phases in a Pickering emulsion system. The emulsions were transferred into a Teflon tube (diameter = 12.7 mm). A thermal polymerization reaction was carried out by heating the emulsions at 90 °C for 2 h. The prepared foams were dried at 50 °C overnight.

The micromorphology of foams was assessed by SEM (F16502, PHENOM, Eindhoven, Netherlands). Then, the area and size of pores in the range of 540 µm × 540 µm were measured by image J, and the specific surface of unit volume (δ) of foams was calculated according to Equation (1).

$$\delta = A/(L \cdot W \cdot d) \quad (1)$$

where A is the area of pores; L and W are the length and width of foams; and d is the average diameter of pores.

The thermal conductivity (λ) of foams was calculated according to Equation (2). The foam was cut into a sheet 12.7 mm in diameter and 1 mm thick; then, the density (ρ) of the sample was measured. The specific heat capacity (Cp) of each foam was measured by differential scanning calorimetry (DSC, 3500 Sirius, NETZSCH-Gerätebau GmbH, Selb, Germany). The initial and termination temperature of system were set to 20 and 40 °C, respectively, and the specific heat capacity of the foam was measured at 35 °C. The thermal diffusivity (α) of foams was measured by laser thermal conductivity meter (LFA467, NETZSCH-Gerätebau GmbH, Selb, Germany) at 35 °C.

$$\lambda = \alpha \cdot Cp \cdot \rho \quad (2)$$

3. Results and Discussion

3.1. Morphology and Properties of Stabilized Particles

To study the assembly structure of particles, TEM images of particles are shown in Figure 1. It can be seen that TOCNC were uniformly dispersed, short rod-shaped fibers. In order to improve the hydrophobicity of TOCNC, TCN particles were prepared by combining nisin with TOCNC. In the TEM images of TCN particles, TOCNC were entangled with each other and aggregated due to the electrostatic attraction with nisin. On the basis of TCN particles, WPU was introduced to prepare TCNW composite particles. In TCNW particles, WPU particles were adsorbed on the surface of aggregated TOCNC.

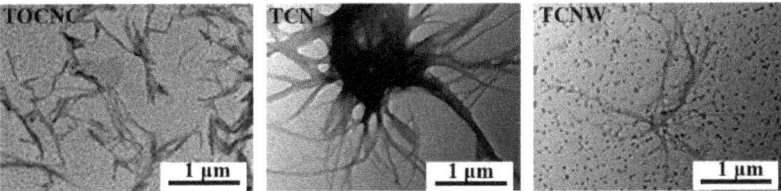

Figure 1. TEM images of TOCNC, TCN and TCNW particles.

Generally, the efficacy of particles to stabilize the bubbles can be assessed by their ability to decrease interfacial tension. As shown in Figure 2a, the interfacial tension of the TOCNC suspension was 78.2 mN/m, which was obviously higher than those of other suspensions. The abundant hydroxyl groups on TOCNC provide hydrophilicity to make it have a less wettability, so it barely had the ability to reduce the interfacial tension. On the other hand, compared with TOCNC suspension, the interfacial tension of the TCN suspension decreased to 54.4 mN/m due to the cationic nisin being electrostatically adsorbed on the anionic TOCNC to improve the wettability of TOCNC at the gas–liquid interface [34]. The introduction of WPU made the interfacial tension of the TCNW suspension further decrease to 49.3 mN/m; the presence of hydrophobic and hydrophilic functional groups in WPU gives it a strong amphiphilic property at the gas–liquid interface. The suspension foaming heights at different times and foaming-effect pictures are shown in Figure 2b. The TOCNC suspension can only be seen to have had a very thin foam layer after shear foaming, TOCNC alone barely had foaming ability. The foaming effects of TCN and TCNW suspensions were obvious after shearing; they both had a thick foam layer, and the height of the foam layer decreased slowly with the time. In addition, the foaming height of the TCNW suspension decreased more slowly compared to that of the TCN suspension. It can be seen that both TCN and TCNW particles foam up better than TOCNC alone, and TCNW particles had the better foam-stabilization effect.

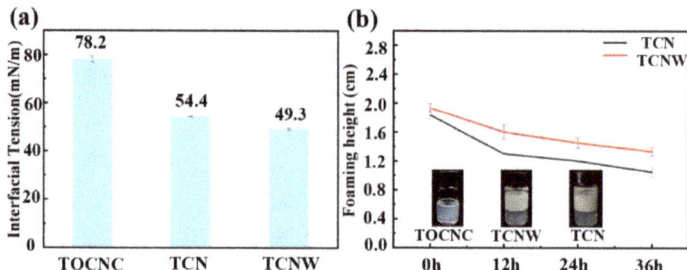

Figure 2. (a) Interfacial tensions of particle suspensions. (b) The suspension foaming heights at different times and foaming effect pictures.

As shown in the optical microscope images of bubbles in Figure 3, the TCN-particles-stabilized bubbles were fewer, and their shape was irregular, whereas the TCNW-particles-stabilized bubbles were more, and they were small. A possible explanation for this discrepancy may be that the adsorption layer formed by TCN particles on the surface of a bubble is thin, meaning it cannot protect the bubble effectively, resulting in the bubbles being deformed under the influence of pressure. The addition of WPU or TCNW particles can form a denser particle shell at the gas–liquid interface to stabilize bubbles. This indicates that nisin and WPU have a synergistic effect on the stabilization of bubbles, and this combination could be used for stabilizing a w/o Pickering emulsion.

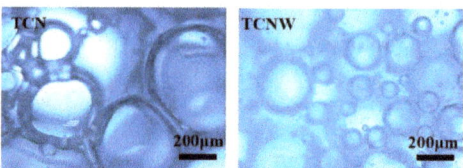

Figure 3. Optical micrographs of TCN- and TCNW-stabilized bubbles.

3.2. Stability of Pickering Emulsions

The stability and droplets size of the Pickering emulsion determine the pore structure and properties of the prepared foam. The micromorphology of emulsions was observed by optical microscope. As Figure 4 shows, the droplets of TOCNC emulsion were less, and the range of droplet size distribution was wide. In the TOCNC emulsion, it can be seen that the droplet size of the TOCNC emulsion was larger than those of TCN and TCNW emulsions, but all three emulsions were stable without creaming. We believe the large droplets have a strong tendency to aggregate with each other, which may result in the instability of the emulsion [35]. In a Pickering emulsion system, the size of the droplets is highly related to the amount of particle stabilizer on the surface of each droplet. In a TOCNC emulsion, due to the low interfacial wettability of TOCNC particles, the surface of droplets cannot be covered completely by particles, so the stability of the TOCNC emulsion was less than that of the other emulsions, thereby resulting in the formation of larger droplets in the emulsion. On the contrary, with the introduction of nisin and WPU, the interfacial wettability of TOCNC was improved, and the obtained composite particles could adsorb on the droplet's surface to form a dense adsorption layer, thereby allowing the formation of smaller droplets. The TCNW emulsion had a narrower range of droplet size distribution, which provides a favorable condition for the preparation of porous foam.

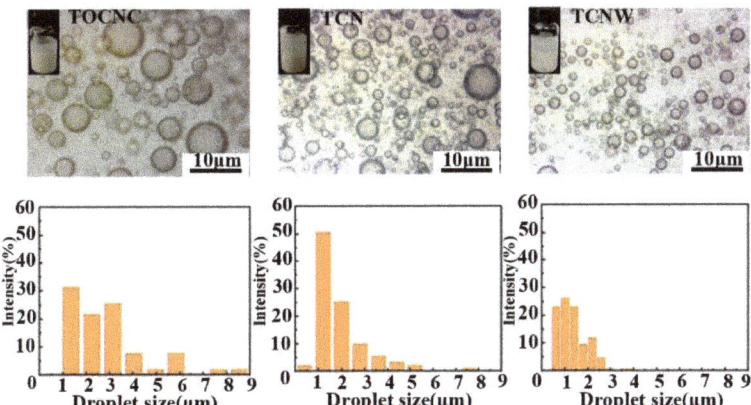

Figure 4. Optical micrographs and droplet size distribution of emulsions.

The stability of emulsions can be determined by rheological testing. The modulus variation of emulsions within the LVR is shown in Figure 5. Typically, emulsions with a wide LVR have a larger limit deformation value, which indicates the emulsion structure has better stability [36]. Compared to the TOCNC emulsion, the limit deformation values of TCN and TCNW emulsions increased by 52% and 54%, respectively, indicating that they both had better stability. Within the LVR, three emulsions showed that G′ was lower than G″; the rheological behavior of emulsions was mainly viscous. The G″ of all emulsions were similar, though the G′ of TCN and TCNW emulsions had increased compared to that of TOCNC. A possible mechanism may be that G″ depends mainly on the continuous phase of emulsion, and the three emulsions have the same continuous phase of AESO; thus, they have similar G″ values. On the other hand, G′ is related to the droplet distribution in the dispersed phase, so uniform droplet distribution resulted in higher G′ for TCN and TCNW emulsions compared to the TOCNC one. In theory, the emulsion's stability can be evaluated at the macroscopic level (bulk emulsion stability), mesoscopic level (droplet size and distribution) and microscopic level (interfacial shear rheology). From the microscopic point of view, the introduction of hydrophobic nisin and soft colloidal WPU facilitates the interface adsorption and elasticity of composite particles, benefiting the stability of the emulsion [36].

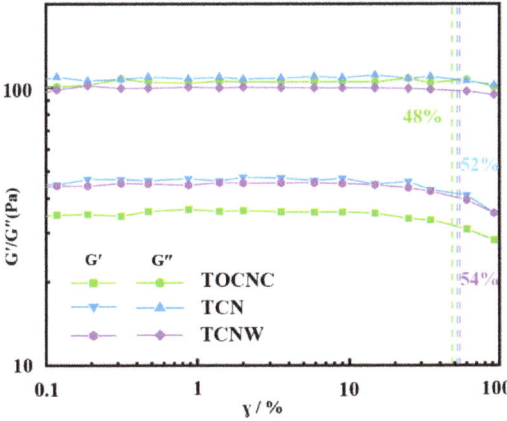

Figure 5. The changes of the storage moduli (G′) and loss moduli (G″) of emulsions.

The stability of a Pickering emulsion is highly related to the adsorption behavior of the particles at the oil-water interface. In this regard, we have the following assumption. As Figure 6 shows, generally, due to the low interfacial wettability of TOCNC, resulting in its lesser adsorption at the oil–water interface, TOCNC particles remain individually dispersed on the droplet surface due to the electrostatic repulsion between the particles. In TCN particles, TOCNC with nisin patches on the surface bridge with adjacent fibers by electrostatic interactions, and the surface wettability of TOCNC is improved, so the adsorption of particles at the oil-water interface has increased, which in turn improves the stability of the emulsion. Unlike rigid particles, WPU as a soft colloid can give Pickering emulsions a deformable interface [28]. In TCNW particles, the introduction of WPU allows for a closer fiber alignment and forms a soft protective barrier on the surfaces of droplets, thereby the emulsion can be stabilized more efficiently. In a Pickering emulsion, we believe the thermal insulation performance of a foam is highly related to the stability of the emulsion. The stability of the emulsion tends to determine the pore structure of the prepared foam, thereby affecting the thermal insulation properties of the foam.

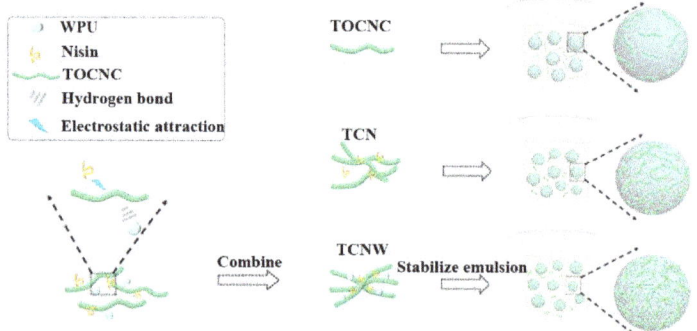

Figure 6. Adsorption of TOCNC, TCN and TCNW particles on emulsion droplets.

3.3. Micromorphology and Thermal Insulation Properties of Foams

After confirming that the physically modified TOCNC can stabilize w/o Pickering emulsion, porous foams were prepared using the Pickering emulsion as a template. The SEM images and pore size distributions of foams are shown in Figure 7. Typical pore structure for the prepared foams can be observed. However, compared to emulsion droplets, the pores were significantly larger. A probable hypothesis is that the loss of thermal stability and coalescence of the emulsion under the thermopolymerization process caused this [33]. In the SEM images of Figure 7, TOCNC foam has a less porous structure than other foams with a non-uniform pore size varying from 4.42 to 90.3 μm. The pores of TCN and TCNW foam were much larger, and TCN and TCNW particles could be observed on the pore wall. By calculating the specific surface of unit volume, the specific surface area of TOCNC foam was found to be 0.0235 $\mu m^2/\mu m^3$. On the other hand, due to the decrease in pore size and increase in the number of pores, the specific surface areas of TCN and TCNW foams increased to 0.0439 and 0.0612 $\mu m^2/\mu m^3$, respectively. In porous materials, a large specific surface area means that the foam has more pore walls, which means that the pores have a stronger capacity to absorb thermal radiation [37].

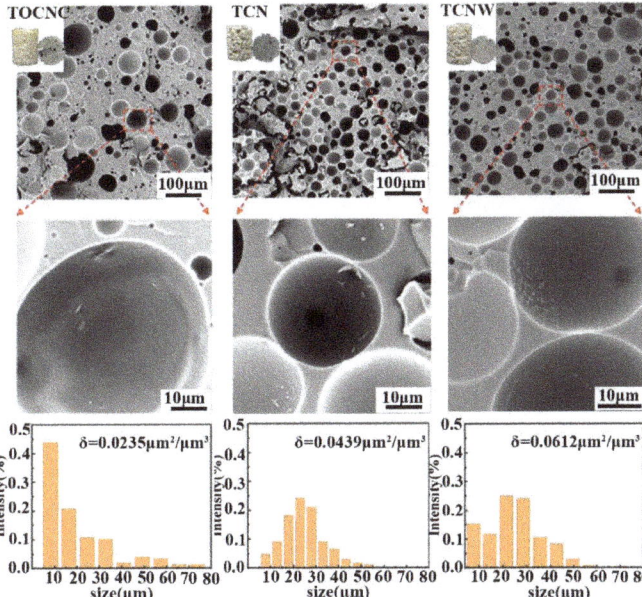

Figure 7. SEM images, specific surface of unit volumes and pore size distributions of foams.

The thermal insulation performance of foam can be characterized by thermal conductivity. Figure 8a shows the thermal conductivity of foams prepared by different particle-stabilized Pickering emulsions. The thermal conductivity of the TOCNC foam was 3.92 W/m·K; and the thermal conductivities of TCN and TCNW foams were significantly lower, at 1.21 and 0.33 W/m·K, respectively. It can be seen that the thermal conductivity of foams decreased with the increase in the number of pores and the decrease in pore size. This phenomenon can be explained by the heat transfer mechanism of foam. The conduction path of heat in the foam is shown in Figure 8b. Generally, the heat conduction in porous material can be divided into gas conduction, solid conduction, radiation conduction and thermal convection within pores [38]. Convection is usually negligible for foams with pore diameters less than 3 mm. The overall thermal conductivity can be expressed as a superposition of the mechanisms, as follows in Equation (3) [39]:

$$\lambda_t = \lambda_g + \lambda_s + \lambda_r \tag{3}$$

where λ_t is the overall thermal conductivity, λ_g is the gas thermal conductivity, λ_s is the solid thermal conductivity and λ_r is the conductivity by radiation.

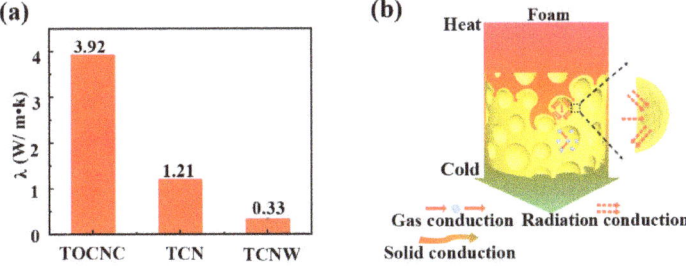

Figure 8. (a) Thermal conductivities of foams; (b) heat conduction in foams.

The above three thermal conductivities can be decreased by reducing the size of pore. Among them, gas conduction is the main mechanism of heat transfer in foam materials. Additionally, gas thermal conductivity can be calculated by Equation (4):

$$\lambda_g = \frac{\in_{VF}}{1 + 2\beta \cdot K_n} k_g \quad (4)$$

where β is the efficiency of energy transfer between the gas molecule and the pore wall; k_g is the conductivity of the gas; $\in_{VF}$ is the void ratio of the foam; K_n is the Knudsen number, which is the radio of mean free path of gas molecules (l_{mean}) to the pore size (φ_c), and can be calculated by Equation (5) [40]:

$$K_n = \frac{l_{mean}}{\varphi_c} \quad (5)$$

When the pore size is equal to or smaller than the average free range of the gas, K_n increases, and the energy transfer through the gas molecules is reduced.

It can be known that Gas conduction is related to the pore size and the average free range of air molecules [9]. In porous foam, the gas thermal conductivity will decrease with the reduction of pore size due to the well-known Knudsen effect [41,42]. On the other hand, with an increase in pores, the foam has more pore walls and a larger specific surface area; the capacity of the foam to absorb, reflect and scatter has enhanced, thereby reducing the thermal conductivity of the foam [43,44].

4. Conclusions

In this study, bio-based foams were obtained by polymerization of an AESO Pickering emulsion which was stabilized by physically modifying TOCNC. After TOCNC was physically modified using WPU and nisin, the interfacial tension of particles was reduced from 78.2 to 49.3 mN/m, and foaming experiments confirmed that the introduction of WPU and nisin has a synergistic effect in enhancing the stability of gas–liquid interface. Rheological tests of the emulsions showed that the stability of composite-particles-prepared emulsions was improved. In the SEM images, the foams prepared by TCN and TCNW particles had a more uniform pore structure, and the specific surface area of TCNW foam increased to 0.0612 μm^2/μm^3. Due to the increases in pores and specific surface area, the thermal conductivity of the TCNW foam was reduced to 0.33 W/M·K compared to the TOCNC. In this study, WPU and nisin were used to physically modify TOCNC to improve the stability of emulsion, thereby obtaining the foam with a uniform pore structure, and the thermal conductivity of the foam was reduced. The strategy developed in this study is expected to be applied to bio-based porous materials in the packaging field.

Author Contributions: Conceptualization, P.L.; formal analysis, Y.L.; resources, H.Z., K.L. and Y.L.; data curation, Y.D. and K.L.; writing—original draft preparation, Y.C.; writing—review and editing, P.L.; visualization, H.Z.; Methodology, M.W. All authors have read and agreed to the published version of the manuscript.

Funding: This research was funded by National Natural Science Foundation of China (No. 22068005) (No. 22168006) and Autonomous District College Student Innovation and Entrepreneurship Training Program (S202210593300).

Institutional Review Board Statement: Not applicable.

Informed Consent Statement: Not applicable.

Data Availability Statement: Not applicable.

Conflicts of Interest: The authors declare no conflict of interest.

References

1. Gama, N.V.; Ferreira, A.; Barros-Timmons, A. Polyurethane Foams: Past, Present, and Future. *Materials* **2018**, *11*, 1841. [CrossRef]
2. Hassan, M.M.; Tucker, N.; Le Guen, M.J. Thermal, mechanical and viscoelastic properties of citric acid-crosslinked starch/cellulose composite foams. *Carbohydr. Polym.* **2020**, *230*, 115675. [CrossRef]
3. Qiu, J.F.; Zhang, M.Q.; Rong, M.Z.; Wu, S.P.; Karger-Kocsis, J. Rigid bio-foam plastics with intrinsic flame retardancy derived from soybean oil. *J. Mater. Chem. A* **2013**, *1*, 2533–2542. [CrossRef]
4. Luo, J.Y.; Zhu, M.; Wang, L.Z.; Zhou, H.F.; Wen, B.Y.; Wang, X.D.; Zhang, Y.X. CO_2-based fabrication of biobased and biodegradable poly (3-hydroxybutyrate-co-3-hydroxyvalerate)/graphene nanoplates nanocomposite foams: Toward EMI shielding application. *Polymer* **2022**, *253*, 125034. [CrossRef]
5. Capron, I.; Rojas, O.J.; Bordes, R. Behavior of nanocelluloses at interfaces. *Curr. Opin. Colloid Interface Sci.* **2017**, *29*, 83–95. [CrossRef]
6. Jiang, H.; Sheng, Y.F.; Ngai, T. Pickering emulsions: Versatility of colloidal particles and recent applications. *Curr. Opin. Colloid Interface Sci.* **2020**, *49*, 1–15. [CrossRef]
7. Ahankari, S.; Paliwal, P.; Subhedar, A.; Kargarzadeh, H. Recent Developments in Nanocellulose-Based Aerogels in Thermal Applications: A Review. *ACS Nano* **2021**, *15*, 3849–3874. [CrossRef]
8. Ahankari, S.S.; Subhedar, A.R.; Bhadauria, S.S.; Dufresne, A. Nanocellulose in food packaging: A review. *Carbohydr. Polym.* **2021**, *255*, 117479. [CrossRef]
9. Apostolopoulou-Kalkavoura, V.; Munier, P.; Bergstrom, L. Thermally Insulating Nanocellulose-Based Materials. *Adv. Mater.* **2021**, *33*, e2001839. [CrossRef]
10. Wang, D.; Peng, H.Y.; Yu, B.; Zhou, K.Q.; Pan, H.F.; Zhang, L.P.; Li, M.; Liu, M.M.; Tian, A.L.; Fu, S.H. Biomimetic structural cellulose nanofiber aerogels with exceptional mechanical, flame-retardant and thermal-insulating properties. *Chem. Eng. J.* **2020**, *389*, 124449. [CrossRef]
11. Brodin, F.W.; Gregersen, O.W.; Syverud, K. Cellulose nanofibrils: Challenges and possibilities as a paper additive or coating material-A review. *Nord. Pulp Pap. Res. J.* **2014**, *29*, 156–166. [CrossRef]
12. Curvello, R.; Raghuwanshi, V.S.; Garnier, G. Engineering nanocellulose hydrogels for biomedical applications. *Adv. Colloid Interface Sci.* **2019**, *267*, 47–61. [CrossRef] [PubMed]
13. Costa, C.; Medronho, B.; Filipe, A.; Mira, I.; Lindman, B.; Edlund, H.; Norgren, M. Emulsion Formation and Stabilization by Biomolecules: The Leading Role of Cellulose. *Polymers* **2019**, *11*, 1570. [CrossRef] [PubMed]
14. Velasquez-Cock, J.; Serpa, A.M.; Gomez-Hoyos, C.; Ganan, P.; Romero-Saez, M.; Velez, L.M.; Correa-Hincapie, N.; Zuluaga, R. Influence of a Non-Ionic Surfactant in the Microstructure and Rheology of a Pickering Emulsion Stabilized by Cellulose Nanofibrils. *Polymers* **2021**, *13*, 3625. [CrossRef]
15. Ataeian, P.; Shi, Q.; Ioannidis, M.; Tam, K.C. Effect of hydrophobic modification of cellulose nanocrystal (CNC) and salt addition on Pickering emulsions undergoing phase-transition. *Carbohydr. Polym. Technol. Appl.* **2022**, *3*, 100201. [CrossRef]
16. Qao, M.; Yang, X.; Zhu, Y.; Guerin, G.; Zhang, S. Ultralight Aerogels with Hierarchical Porous Structures Prepared from Cellulose Nanocrystal Stabilized Pickering High Internal Phase Emulsions. *Langmuir* **2020**, *36*, 6421–6428. [CrossRef]
17. Tasset, S.; Cathala, B.; Bizot, H.; Capron, I. Versatile cellular foams derived from CNC-stabilized Pickering emulsions. *Rsc Adv.* **2014**, *4*, 893–898. [CrossRef]
18. He, Y.; Li, S.; Zhou, L.; Wei, C.; Yu, C.; Chen, Y.; Liu, H. Cellulose nanofibrils-based hybrid foam generated from Pickering emulsion toward high-performance microwave absorption. *Carbohydr. Polym.* **2021**, *255*, 117333. [CrossRef]
19. Tian, X.W.; Ge, X.H.; Guo, M.Y.; Ma, J.X.; Meng, Z.Q.; Lu, P. An antimicrobial bio-based polymer foam from ZnO-stabilised pickering emulsion templated polymerisation. *J. Mater. Sci.* **2021**, *56*, 1643–1657. [CrossRef]
20. Bai, L.; Huan, S.Q.; Zhu, Y.; Chu, G.; McClements, D.J.; Rojas, O.J. Recent Advances in Food Emulsions and Engineering Foodstuffs Using Plant-Based Nanocelluloses. *Annu. Rev. Food Sci. Technol.* **2021**, *12*, 383–406. [CrossRef]
21. Wu, Z.H.; Xu, J.; Gong, J.; Li, J.; Mo, L.H. Preparation, characterization and acetylation of cellulose nanocrystal allomorphs. *Cellulose* **2018**, *25*, 4905–4918. [CrossRef]
22. Hu, Z.; Ballinger, S.; Pelton, R.; Cranston, E.D. Surfactant-enhanced cellulose nanocrystal Pickering emulsions. *J. Colloid Interface Sci.* **2015**, *439*, 139–148. [CrossRef] [PubMed]
23. Lee, K.-Y.; Blaker, J.J.; Murakami, R.; Heng, J.Y.Y.; Bismarck, A. Phase Behavior of Medium and High Internal Phase Water-in-Oil Emulsions Stabilized Solely by Hydrophobized Bacterial Cellulose Nanofibrils. *Langmuir* **2014**, *30*, 452–460. [CrossRef] [PubMed]
24. Wu, H.; Teng, C.; Liu, B.; Tian, H.; Wang, J. Characterization and long term antimicrobial activity of the nisin anchored cellulose films. *Int. J. Biol. Macromol.* **2018**, *113*, 487–493. [CrossRef] [PubMed]
25. Zhao, H.; Gao, W.-C.; Li, Q.; Khan, M.R.; Hu, G.-H.; Liu, Y.; Wu, W.; Huang, C.-X.; Li, R.K.Y. Recent advances in superhydrophobic polyurethane: Preparations and applications. *Adv. Colloid Interface Sci.* **2022**, *303*, 102644. [CrossRef] [PubMed]
26. Cheng, B.-X.; Lu, C.-C.; Li, Q.; Zhao, S.-Q.; Bi, C.-S.; Wu, W.; Huang, C.-X.; Zhao, H. Preparation and Properties of Self-healing Triboelectric Nanogenerator Based on Waterborne Polyurethane Containing Diels–Alder Bonds. *J. Polym. Environ.* **2022**, *30*, 5252–5262. [CrossRef]
27. Zhao, H.; Li, K.-C.; Wu, W.; Li, Q.; Jiang, Y.; Cheng, B.-X.; Huang, C.-X.; Li, H.-N. Microstructure and viscoelastic behavior of waterborne polyurethane/cellulose nanofiber nanocomposite. *J. Ind. Eng. Chem.* **2022**, *110*, 150–157. [CrossRef]

28. Wu, J.; Guan, X.; Wang, C.; Ngai, T.; Lin, W. pH-Responsive Pickering high internal phase emulsions stabilized by Waterborne polyurethane. *J. Colloid Interface Sci.* **2022**, *610*, 994–1004. [CrossRef]
29. Liu, W.; Fei, M.-E.; Ban, Y.; Jia, A.; Qiu, R. Preparation and Evaluation of Green Composites from Microcrystalline Cellulose and a Soybean-Oil Derivative. *Polymers* **2017**, *9*, 541. [CrossRef]
30. Bonnaillie, L.M.; Wool, R.P. Thermosetting foam with a high bio-based content from acrylated epoxidized soybean oil and carbon dioxide. *J. Appl. Polym. Sci.* **2007**, *105*, 1042–1052. [CrossRef]
31. Yang, Y.; Zhang, M.; Sha, L.; Lu, P.; Wu, M. "Bottom-Up" Assembly of Nanocellulose Microgels as Stabilizer for Pickering Foam Forming. *Biomacromolecules* **2021**, *22*, 3960–3970. [CrossRef] [PubMed]
32. Lu, P.; Zhao, H.; Zhang, M.; Bi, X.; Ge, X.; Wu, M. Thermal insulation and antibacterial foam templated from bagasse nanocellulose/nisin complex stabilized Pickering emulsion. *Colloids Surf. B. Biointerfaces* **2022**, *220*, 112881. [CrossRef]
33. Lu, P.; Guo, M.; Yang, Y.; Wu, M. Nanocellulose Stabilized Pickering Emulsion Templating for Thermosetting AESO Nanocomposite Foams. *Polymers* **2018**, *10*, 1111. [CrossRef] [PubMed]
34. Zhan, F.; Youssef, M.; Li, J.; Li, B. Beyond particle stabilization of emulsions and foams: Proteins in liquid-liquid and liquid-gas interfaces. *Adv. Colloid Interface Sci.* **2022**, *308*, 102743. [CrossRef] [PubMed]
35. Tcholakova, S.; Denkov, N.D.; Sidzhakova, D.; Ivanov, I.B.; Campbell, B. Interrelation between Drop Size and Protein Adsorption at Various Emulsification Conditions. *Langmuir* **2003**, *19*, 5640–5649. [CrossRef]
36. Jia, Y.; Kong, L.; Zhang, B.; Fu, X.; Huang, Q. Fabrication and characterization of Pickering high internal phase emulsions stabilized by debranched starch-capric acid complex nanoparticles. *Int. J. Biol. Macromol.* **2022**, *207*, 791–800. [CrossRef]
37. Baillis, D.; Coquard, R. Radiative and Conductive Thermal Properties of Foams. *Cellular and Porous Materials.* **2008**, 343–384. [CrossRef]
38. Zhang, H.; Zhang, G.; Gao, Q.; Tang, M.; Ma, Z.; Qin, J.; Wang, M.; Kim, J.-K. Multifunctional microcellular PVDF/Ni-chains composite foams with enhanced electromagnetic interference shielding and superior thermal insulation performance. *Chem. Eng. J.* **2020**, *379*, 122304. [CrossRef]
39. Hasanzadeh, R.; Azdast, T.; Doniavi, A.; Lee, R.E. Thermal-insulation performance of low density polyethylene (LDPE) foams: Comparison between two radiation thermal conductivity models. *Polyolefins J.* **2019**, *6*, 13–21. [CrossRef]
40. Hasanzadeh, R.; Azdast, T.; Doniavi, A.; Rostami, M. A prediction model using response surface methodology based on cell size and foam density to predict thermal conductivity of polystyrene foams. *Heat Mass Transfer.* **2019**, *55*, 2845–2855. [CrossRef]
41. Hasanzadeh, R.; Azdast, T.; Doniavi, A.; Lee, R.E. Multi-objective optimization of heat transfer mechanisms of microcellular polymeric foams from thermal-insulation point of view. *Therm. Sci. Eng. Prog.* **2019**, *9*, 21–29. [CrossRef]
42. Hasanzadeh, R.; Azdast, T.; Doniavi, A. Thermal Conductivity of Low-Density Polyethylene Foams Part II: Deep Investigation using Response Surface Methodology. *J. Therm. Sci.* **2020**, *29*, 159–168. [CrossRef]
43. Hasanzadeh, R.; Fathi, S.; Azdast, T.; Rostami, M. Theoretical Investigation and Optimization of Radiation Thermal Conduction of Thermal-Insulation Polyolefin Foams. *J. Iran. J. Mater. Sci. Eng.* **2020**, *17*, 58–65. [CrossRef]
44. Wang, G.; Zhao, J.; Wang, G.; Mark, L.H.; Park, C.B.; Zhao, G. Low-density and structure-tunable microcellular PMMA foams with improved thermal-insulation and compressive mechanical properties. *Eur. Polym. J.* **2017**, *95*, 382–393. [CrossRef]

Article

Study on Degradation of 1,2,4-TrCB by Sugarcane Cellulose-TiO₂ Carrier in an Intimate Coupling of Photocatalysis and Biodegradation System

Zhenqi Zhou [1,†], Chunlin Jiao [1,†], Yinna Liang [2], Ang Du [1], Jiaming Zhang [1], Jianhua Xiong [1,*], Guoning Chen [3], Hongxiang Zhu [2,4] and Lihai Lu [3]

[1] School of Resources, Environment and Materials, Guangxi University, Nanning 530004, China
[2] College of Light Industry and Food Engineering, Guangxi University, Nanning 530004, China
[3] Guangxi Bossco Environmental Protection Technology Co., Ltd., Nanning 530007, China
[4] Guangxi Key Laboratory of Clean Pulp & Papermaking and Pollution Control, Nanning 530004, China
* Correspondence: xjh@gxu.edu.cn
† These authors contributed equally to this work.

Abstract: 1,2,4 trichlorobenzene (1,2,4-TrCB) is a persistent organic pollutant with chemical stability, biological toxicity, and durability, which has a significant adverse impact on the ecological environment and human health. In order to solve the pollution problem, bagasse cellulose is used as the basic framework and nano TiO₂ is used as the photocatalyst to prepare composite carriers with excellent performance. Based on this, an intimate coupling of photocatalysis and biodegradation (ICPB) system combining photocatalysis and microorganisms is constructed. We use the combined technology for the first time to deal with the pollution problem of 1,2,4-TrCB. The biofilm in the composite carrier can decompose the photocatalytic products so that the removal rate of 1,2,4-TrCB is 68.01%, which is 14.81% higher than those of biodegradation or photocatalysis alone, and the mineralization rate is 50.30%, which is 11.50% higher than that of photocatalysis alone. The degradation pathways and mechanisms of 1,2,4-TrCB are explored, which provide a theoretical basis and potential application for the efficient degradation of 1,2,4-TrCB and other refractory organics by the ICPB system.

Keywords: sugarcane cellulose; photocatalysis; microorganism; ICPB; 1,2,4-TrCB

1. Introduction

1,2,4 trichlorobenzene(1,2,4-TrCB) is a typical persistent organic pollutant with characteristic stability and biological toxicity that degrades slowly in its natural state. It is widely found in soil, groundwater, wastewater, and agricultural crops and easily spreads in the environment, which can cause serious harm to the ecological environment and human health [1,2]. Physical and chemical treatments for 1,2,4-TrCB contamination include adsorption, pyrolysis, oxidation, and electrochemistry. However, there are still some deficiencies such as incomplete degradation, secondary pollution, and requirements for operating costs and conditions [3]. Although the microbial degradation of organic pollutants is simple, economical, and free of secondary pollutants, its efficiency in the degradation process is limited [4,5]. At present, there are many studies on single repair techniques for 1,2,4-TrCB, but some defects exist and studies on combinations of multiple techniques are only preliminary [6].

Pollution by 1,2,4-TrCB urgently needs to be effectively solved to improve the ecological environment and alleviate the risk to human health. Therefore, for 1,2,4-TrCB, this paper uses a joint technology to fix the problem, hoping to explore a practical and effective remediation technology, which is of great significance for solving the pollution problem of refractory organic matter such as this pollutant.

The intimate coupling of photocatalysis and biodegradation (ICPB), a new method of water pollution control introduced by Rittman et al. [7] in 2008, combines photocatalytic oxidation and microbial degradation into one reactor and ensures that both are involved simultaneously in the degradation of pollutants. This method has shown superior degradation and mineralization efficiency for persistent pollutants such as phenol [8], tetracycline [9], and chlorophenol [10]. The practical application of ICPB is limited by a low bacterial load density and low microbial activity. At the same time, it also still has issues in terms of reliability, economy, and universality of actual wastewater treatment plants so this technology has not been applied in practice [11,12]. However, the related research is extensive and in-depth and it is believed that ICPB technology has good application potential.

At present, the composite carriers used in the ICPB system include polyurethane sponge carriers, porous ceramics, and cellulose. Because of its low density, polyurethane sponge is conducive to uniform diffusion with water flow and has good adsorbability, but it is difficult to maintain the stability of biofilm after repeated use [13]. Compared with polyurethane sponge carriers, ceramic carriers can maintain the stability of biofilm even after repeated use. However, due to its high density, it is difficult to ensure that it can operate with water flow so the system's efficiency cannot be guaranteed [14]. Cellulose has excellent adsorbability and can immobilize catalysts and biofilm. At the same time, it is a bio-friendly material and will not produce secondary pollution.

Therefore, in this paper, sugarcane cellulose was used as the basic raw material to prepare an environmentally friendly photocatalyst carrier. The ICPB system was used for the first time in the degradation of 1,2,4-TrCB, which broadened the combined technology to the treatment of this pollutant. We investigated 1,2,4-TrCB degradation by comparing the ICPB with photocatalysis and biodegradation individually. Free radical capture experiments were carried out and the possible degradation pathway and responses to the biofilm were analyzed. The dominant microorganisms degrading 1,2,4-TrCB were investigated to illuminate the mechanism of the ICPB, providing a possible theoretical basis and practical method for the efficient degradation of 1,2,4-TrCB and other refractory organic pollutants.

2. Material and Methods

2.1. Preparing Materials

A cellulose carrier was prepared of bagasse cellulose, absorbent cotton, and sodium sulfate (Na_2SO_4) in a zinc chloride ($ZnCl_2$) solution. After it was solidified in deionized water and freeze-dried, a carrier was obtained with a large number of pores. The specific process was as follows: (1) An amount of 1 g of cellulose was added to a 99 g $ZnCl_2$ (70%, wt) solution and after mixing, the mixture (100 g) was stirred at 80 °C for 60 min; (2) Then, 0.6 g absorbent cotton was added into the mixture while stirring for 60 min at 60 °C, followed by the addition of 60 g Na_2SO_4; (3) After stirring for 60 min at 60 °C, the mixture was solidified in deionized water for 2 days and freeze-dried for 2 days at −70 °C. The carriers (5 mm × 5 mm × 5 mm) were obtained. This process is shown in Figure 1.

Figure 1. Preparation of carrier.

The catalyst was loaded onto the carriers via a simple and efficient low-temperature process, which has been previously described. In short, the process was as follows: (1) An amount of 1.5 g of visible light-responsive titanium dioxide (TiO_2) was dissolved in a 15 mL solution of 0.3 g/L defused sodium and the mixture was stirred for 15 min to fully disperse the TiO_2; (2) The prepared cellulose carriers were soaked in the mixed solution for 10 min to fully load the TiO_2 photocatalyst onto the carriers; (3) The carriers were removed and baked at 60 °C for 120 min and then ultrasonically cleaned in deionized water for 5 min, and this process was repeated 3 times. This process is shown in Figure 2. The final sugarcane cellulose-TiO_2 carrier had been prepared.

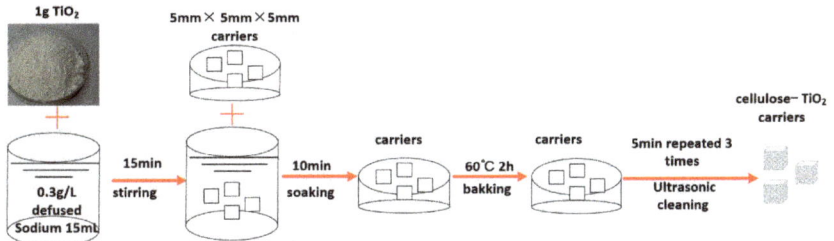

Figure 2. Preparation of cellulose-TiO_2 carrier.

Sugarcane cellulose was obtained from Guangxi Guigang Guitang Co., Ltd., (Guigang, China) and the catalyst (visible light-responsive titanium dioxide) from Liuzhou Rose Nanomaterials Technology Co., Ltd (Liuzhou, China). Zinc chloride was purchased from Tianjin Ou Boke Chemical Products Sales Co., Ltd (Tianjin, China). Sodium sulfate was purchased from Guangdong Guanghua Sci-Tech Co, Ltd., (Shantou, China) and absorbent cotton from Nanchang Leiyi Medical Appliance Co., Ltd (Nanchang, China). Rhodamine B was purchased from Aladdin Reagent Co., Ltd(Shanghai, China). 1,2,4-TrCB was purchased from Macklin. All the chemicals were analytically pure. Activated sludge was derived from the research center of the Guangxi Bossco Environmental Protection Technology Co. in Guangxi province (Nanning, China).

2.2. Methods

(1) Experimental methods

The schematic diagram of the ICPB system is shown in Figure 3. A xenon lamp (XHA250W) was used as the light source set 15 cm away from the quartz baker (500 mL), and a stirrer was used to power the solution at 100 r/min. The initial pH was 5 and the concentrations of 1,2,4-TrCB and dissolved oxygen were 8.0 mg/L and 6.0–7.0 mg/L, respectively. The volume proportion of solution to carriers was 8%. To assess the ICPB degradation ability, the protocols, including photolysis, adsorption, biodegradation, and photocatalysis, were performed at room temperature (~26 °C).

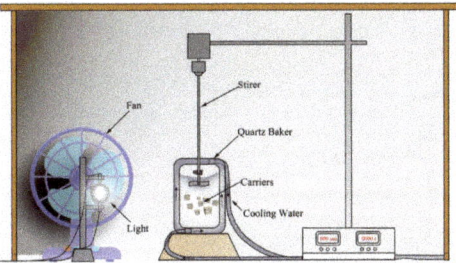

Figure 3. Schematic diagram of equipment.

(2) Analytical methods

The concentrations of 1,2,4-TrCB and the intermediate products were determined by a gas chromatography–mass spectrometer (5975C GC/MSD, Agilent). N-hexane was the extraction agent for the 1,2,4-TrCB and intermediate products. The initial temperature was 50 °C (maintained for 1 min), which was increased in 20 °C/min increments to 200 °C (maintained for 5 min), followed by a 10 °C/min increase to 280 °C (maintained for 1 min). The carrier gas was helium and the column flow rate was 1.0 mL/min. The injector temperature was 250 °C and the detector temperature was 280 °C. The sample of the intermediate products was obtained by concentration to reduce the liquor from 100 mL to 5 mL using rotary evaporators (RE-52A) for detection. We used sodium oxalate (1 mmol/L), tertiary butanol (1 mmol/L), and p-benzoquinone (2 mmol/L) as the trapping agents of the electron hole, hydroxyl radical, and superoxide radical, respectively [15]. A scanning electron microscope (SEM, Hitachi SU8220) was used to observe and analyze the microorganisms in the carrier before and after the experiment. The high-throughput sequencing of different samples containing the initial solution and the solution after biodegradation and ICPB was performed by Shanghai Majorbio Bio-Pharm Technology Co., Ltd. and analyzed on the cloud platform of Majorbio.

3. Results and Discussion
3.1. Degradation of 1,2,4-TrCB in ICPB

Figure 4 shows the degradation of 1,2,4-TrCB in different systems, referring to the reaction processes of adsorption, photodegradation, photocatalytic oxidation, and biodegradation, in addition to those in the ICPB system. The adsorption of the TiO_2-cellulose type carrier for 1,2,4-TrCB was 11.50% after 7 h and adsorption equilibria appeared at the first hour. Therefore, adsorption had little effect on the removal of TCB. In the photolysis experiment, the removal rate of 1,2,4-TrCB was 16.15% after 7 h, indicating that photolysis can degrade 1,2,4-TrCB with low efficiency. This is consistent with the results demonstrated by Wang et al. [16]. In addition, Kozhevnikova et al. [17] found that a 16.2% removal of 1,2,4-TrCB with an initial concentration of 18 mg/L was obtained after 7.5 h under ultraviolet radiation alone (240 W, λ = 240–320 nm). Therefore, high removal was also difficult using ultraviolet radiation alone without a catalyst [18,19]. Compared with ultraviolet radiation, the Xenon lamp used in this study had a lower ability to degrade 1,2,4-TrCB because its wavelength was similar to that of natural light, which has lower energy.

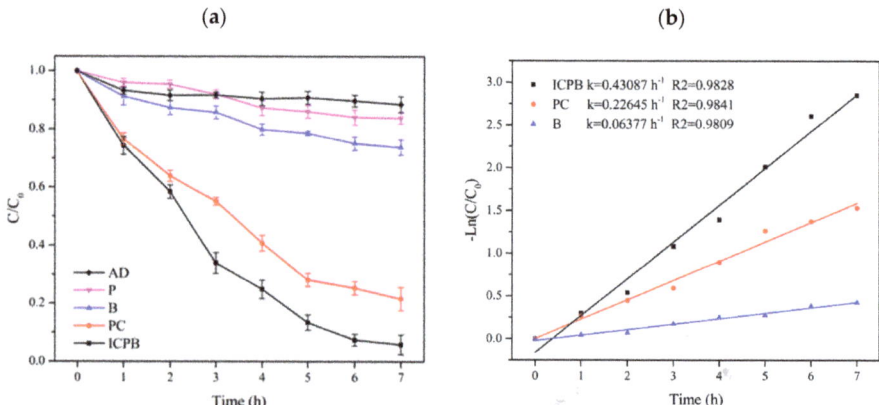

Figure 4. Curves of 1,2,4-TrCB degradation (**a**) and kinetics (**b**) in different protocols. AD—adsorption without light and microorganisms; P—photolysis without carriers; B—biodegradation without light; PC—photocatalysis without microorganisms; ICPB—intimate coupling of biodegradation and photocatalysis.

In the biodegradation system (B) under the condition of microorganisms attached to the carriers without a light source, 26.20% of the substrate was degraded, which was 10.05% higher than that of single photolysis, indicating success in the domestication of the microbial degradation of 1,2,4-TrCB and that the microorganism had a certain ability to degrade 1,2,4-TrCB, as seen in Figure 4. Brunsbach et al. [20] studied indigenous microorganisms in mud soil and found that microorganisms could degrade chlorobenzene only when 1,2,4-TrCB was mixed with low-substituted chlorobenzene. In addition, Dermietzel et al. [21] and Dong et al. [22] also observed the microorganisms' ability to metabolize 1,2,4-TrCB. Therefore, the contribution to 1,2,4-TrCB degradation by microbial sources was found. However, the long duration and low efficiency hinder its widespread application to POPs biodegradation.

Photocatalytic oxidation, an advanced oxidation method with a robust ability to degrade pollutants, is appropriate for decomposing difficult-to-degrade pollutants such as tetracycline [23], chlorobenzene [24], and hexachlorobenzene [25]. In the photocatalysis system (PC) used in this study, the removal rate of 1,2,4-TrCB was up to 79.40% for 7 h and was 53.20% higher than that of biodegradation alone, as shown in Figure 2. It was illustrated that photocatalytic oxidation should have a strong ability to degrade 1,2,4-TrCB. During the photocatalytic reaction, a large number of active groups with strong oxidizing properties, such as pores (h^+) and hydroxyl radicals ($\cdot OH$), oxidize 1,2,4-TrCB into CO_2 and H_2O through chlorine atom substitution and as the ring structure opens gradually.

The removal rate of 1,2,4-TrCB in the ICPB system for 7 h reached 94.21%, which was 68.01% and 14.81% higher than that of the microorganism alone and the photocatalytic oxidation system, respectively (Figure 4). The first-order kinetic rate constant of 1,2,4-TrCB degradation in the ICPB system was 0.43087 h^{-1}, which was 1.9 times and 6.7 times higher than those of photocatalytic and microbial degradation systems, respectively. Kozhevnikova et al. used composite materials to catalyze the degradation of 1,2,4-TrCB and the conversion rate was 32.6% after a reaction for 7.5 h [17]. Song et al. used activated carbon-supported microorganisms to treat 1,2,4-TrCB and the degradation rate reached 48.1% in 23 day [26]. These results indicate that the use of the ICPB composite system proposed in this paper is more effective than photocatalysis or biodegradation alone.

It can be seen that photocatalytic oxidation and microbial degradation in the ICPB system did not inhibit but rather promoted each other, showing a synergistic effect on the removal rate. Considering the ICPB system's performance regarding the degradation of other pollutants, such as phenol [27], 4-chlorophenol [10], and 2,4,6-trichlorophenol [28], the robust ability to degrade POPs appears when closely combining photocatalytic oxidation and microbial degradation.

3.2. Mineralization of 1,2,4-TrCB in ICPB

In order to reveal the mineralization effect on the ICPB system, the content of total organic carbon (TOC) was detected in the 1,2,4-TrCB degradation process using the different systems, as shown in Figure 5. Compared to the photocatalytic oxidation and coupling system, the biodegradation mineralization rate was the lowest. As expected, the power of mineralization from photocatalysis was robust and the removal rate of TOC reached up to 67.8%. However, the effect was still inferior to that of the ICPB system. Moreover, some studies have confirmed that photocatalytic oxidation may cause excessive oxidation of pollutants and does not complete mineralization, accounting for the lower rate compared to that of ICPB. The mineralization rate of 1,2,4-TrCB in the ICPB system was 79.30%, which was 11.50% and 50.30% higher than that of photocatalytic oxidation (PC) and biodegradation (B), respectively, and the first-order kinetic rate constant was increased by 0.46 times and 3.57 times, respectively. Due to the presence of microbial community in the ICPB system, the intermediate products produced by photocatalytic oxidation of 1,2,4-TrCB were relatively low in toxicity and easy to be metabolized and degraded by the microbial community, which reduced the possibility of excessive oxidation to a certain extent. At the same time, the competition between the intermediate and 1,2,4-TrCB for active species was reduced, and degradation and mineralization were promoted. This mechanism has been proved to a certain extent by Wen et al. [29] in the study on 2,4-dinitrotoluene's degradation in the ICPB system: Although the degradation rate of 2,4-dinitrotoluene was close in both the photocatalysis and the ICPB systems, the amount of organic nitrogen transformed into NH_4^+-N, NO_2-N and NO_3-N in the ICPB system highly exceeded that of the photocatalytic oxidation. This indicates that photocatalytic oxidation has a synergistic effect on the microbial community in the system, which promoted the degradation and mineralization of 1,2,4-TrCB.

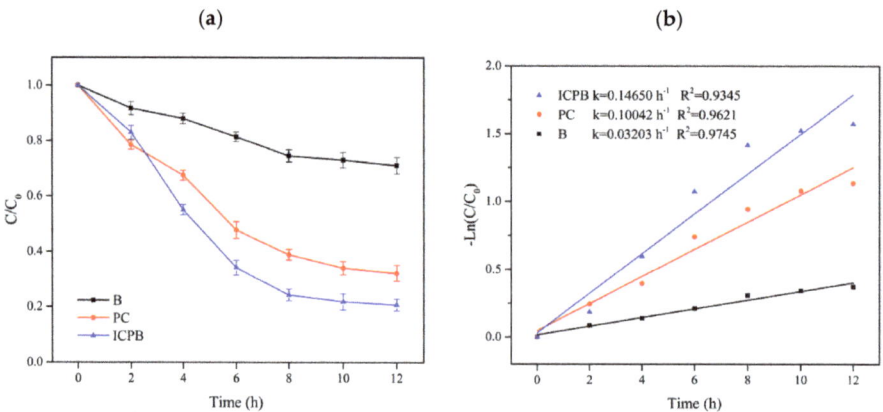

Figure 5. Curves of TOC degradation (**a**) and kinetics (**b**) in B, PC, and ICPB. B—biodegradation without light; PC—photocatalysis without microorganism; ICPB—intimate coupling of biodegradation and photocatalysis.

3.3. Construction of Free Radicals in ICPB

Active species with strong oxidation abilities are the major contributors to pollutant degradation. In this experiment, the effect of electron holes (h^+), hydroxyl radicals ($\cdot OH$), and superoxide radicals ($\cdot O_2-$) on the 1,2,4-TrCB removal rate was explored by adding the trapping agents of the corresponding active species, and the contribution to the different active species by the ICPB system was clearly visible, as shown in Figure 6. The removal rate of 1,2,4-TrCB in the ICPB system decreased from 93.80% to 32.72% when electron holes were trapped. The large decline of 54.46% demonstrated that electron holes play a major role in the degradation of 1,2,4-TrCB. Although they directly oxidized 1,2,4-TrCB [30], electron holes also promoted the production of hydroxyl radicals, as shown in Equations (1) and (2) [31].

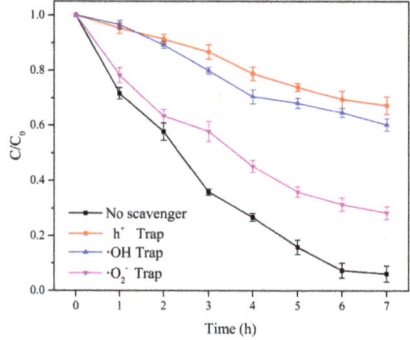

Figure 6. Curves of 1,2,4-TrCB with different trapping agents.

Hydroxyl radicals can oxidize numerous refractory pollutants indiscriminately including 4-chlorophenol [32], hexachlorobenzene, o-dichlorobenzene, and 2,4-dichlorophenoxyacetic acid, among other pollutants. Therefore, hydroxyl radicals play a major role in a liquid-phase photocatalytic reaction [33]. In this experiment, when the capturing agent of hydroxyl radicals was added, the 1,2,4-TrCB removal rate significantly decreased. As seen in Figure 6, the removal rate of 1,2,4-TrCB decreased from 93.80% to 39.77%, indicating that hydroxyl radicals in the ICPB system also significantly contributed to 1,2,4-TrCB degradation.

$$\text{Catalyst} + h\nu \rightarrow e^- + h^+ \tag{1}$$

$$OH^- + h^+ \rightarrow \cdot OH \tag{2}$$

$$e^- + O_2 \rightarrow \cdot O_2^- \tag{3}$$

As described in Equation (3) [31], during the photocatalysis process, the dissolved oxygen in the solution captured electrons and became superoxide radicals, which not only directly participated in pollutant degradation but also promoted the generation of hydroxyl radicals through the protonation process and improved photocatalytic oxidation efficiency. Lichtenberger et al. [34] found that superoxide radicals could oxidize dichlorophenol to chlorophenol. Additionally, Lin et al. [35] proved the oxidization and crack of 1,2,4-TrCB into small molecules under an attack by superoxide radicals. After adding the capture agent of superoxide radicals, the removal rate of 1,2,4-TrCB in the ICPB system decreased from 93.80% to 72.64%, demonstrating that superoxide radicals promoted the degradation of 1,2,4-TrCB, though the effect was weaker than those of hydroxyl radicals and electron holes. Considering the demands of microbial metabolism in the ICPB system, the dissolved oxygen was consumed gradually, which weakened the role of superoxide radicals in this system.

3.4. Degradation Pathway of 1,2,4-TrCB in ICPB

The chromatogram of the 1,2,4-TrCB degradation intermediates in the ICPB system is shown in Figure 7. There were 10 possible intermediate products, as shown in Table 1, including aromatic and chain compounds. The aromatic compounds were, respectively, $C_6H_4Cl_2$ (o-dichlorobenzene, p-dichlorobenzene), $C_{11}H_{10}O_6$ (3,4-bismethoxycarbonyl benzoic acid), $C_{14}H_{22}O$ (2,4-di-dutylphenol), and $C_{16}H_{22}O_4$ (1,2-dicarboxylate -dibutyl benzene) and the chain compounds included $C_{10}H_{22}O$ (tetra-hydrolavandulol), $C_{10}H_{20}O$ (2-decenol), $C_8H_{18}O$ (2-ethylhexanol), and $C_6H_{14}O_2$ (3-Hexyl hydroperoxide), with the functional carboxyl and hydroxyl groups. As described above, photocatalysis significantly contributed to 1,2,4-TrCB degradation, of which the reactions of substitution, addition, and ring opening occurred by the oxidization of active species and produced a series of simple organic compounds [36].

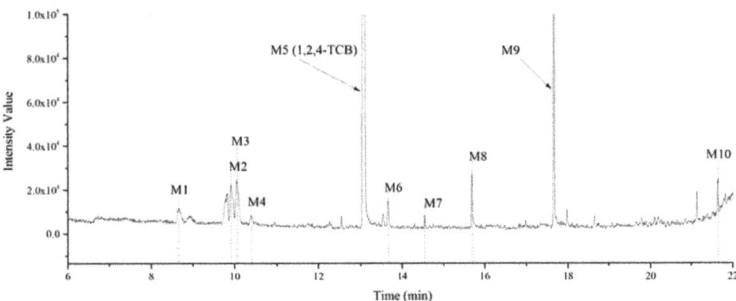

Figure 7. Chromatography of intermediate products of 1,2,4-TrCB degradation. The numbers M1 to M11 are the intermediate products corresponding to Table 1.

Table 1. Intermediate products of 1,2,4-TrCB.

No.	Molecular Formula	m/z	Proposed Molecular	Proposed Structure
M1	$C_6H_{14}O_2$	118.10	3-Hexyl hydroperoxide	
M2	$C_6H_4Cl_2$	145.97	o-Dichlorobenzene	
M3	$C_6H_4Cl_2$	145.97	p-Dichlorobenzene	
M4	$C_8H_{18}O$	130.14	2-Ethylhexanol	
M5	$C_6H_3Cl_3$	179.93	1,2,4-Trichlorobenzene	
M6	$C_{10}H_{20}O$	156.15	2-Decenol	
M7	$C_{10}H_{22}O$	158.17	Tetrahydrolavandulol	
M8	$C_{11}H_{10}O_6$	238.05	3,4-Bis (methoxycarbonyl) benzoic acid	
M9	$C_{14}H_{22}O$	206.17	2,4-di-Butylphenol	
M10	$C_{16}H_{22}O_4$	278.15	1,2-dicarboxylate -dibutyl benzene	

When 1,2,4-TrCB was adsorbed to the surface of the catalyst, numerous opportunities for oxidation reactions occurred between the substrates and the active groups. The experimental investigations showed that electron holes directly reacted with 1,2,4-TrCB to produce dichlorobenzene radical (·DPC) and other substances [17]. Furthermore, 1,2,4-TrCB gradually dechlorinated to produce chlorobenzene with low substituted numbers. Breaking the bond between C and Cl in the second position, first position, and fourth position in order needed progressively more energy [37]. Therefore, para-dichlorobenzene dominated the list of intermediate products, followed by m-dichlorobenzene and ortho-dichlorobenzene, respectively. However, m-dichlorobenzene could not be detected in this study for unclear reasons and should be further explored. For the low-substituted chlorobenzenes or others, chlorophenols or phenolic substances were produced through nucleophilic substitution by the attack of the hydroxyl and superoxide radicals, and even the ring structure broke into the single-stranded structure [38]. The products were further oxidized and decomposed into small molecules and further mineralized into CO_2 and H_2O. Considering what follows in this paper, microflora played an important role in the degradation of 1,2,4-TrCB, though the exact mineralization pathway is unknown.

The general degradation pathway is introduced in Figure 8. Firstly, the chlorobenzenes (M2, M3) with low substitution numbers were produced by oxidation and substitution reactions. Secondly, the chlorobenzenes were further oxidized into phenolic compounds or reacted with others to form ester compounds (M8, M9, M10). Next, the ring structures cracked into the short-chain fatty acids or alcohol (M1, M4, M6, M7) and were finally transformed into CO_2 and H_2O. The results were consistent with the degradation rule of 1,2,4-TrCB and other pollutants' decompositions in the ICPB, which to some extent revealed the degradation process of 1,2,4-TrCB in the ICPB and further deepened the understanding of the operational mechanism of the ICPB system. However, further research is required on different degradation systems and the metabolic process of microorganisms to fully unravel the specific pathway.

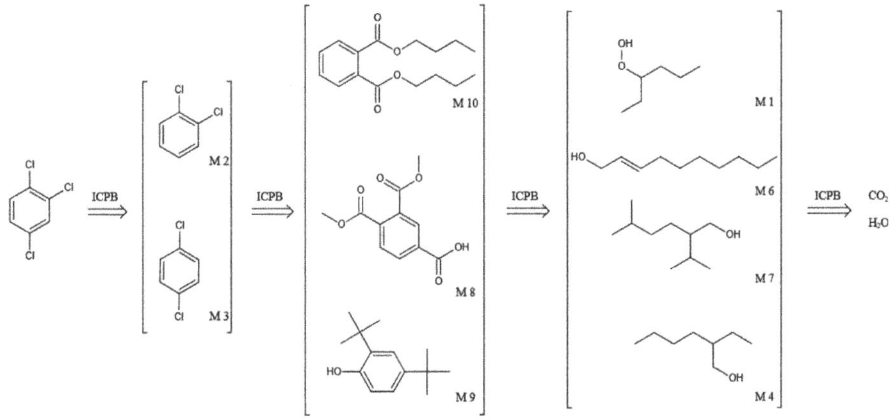

Figure 8. Degradation pathway of 1,2,4-TrCB.

3.5. Microbial Community Response Analysis

3.5.1. SEM Analysis of Microorganisms in Carriers

We wanted to confirm whether the microorganisms were still stably attached to the carriers after the ICPB system had been continuously run for six cycles. Therefore, the carriers were removed for the scanning electron microscope observation in addition to the other carriers that were not included in the reaction as the control group. A random sampling of the carriers revealed abundant microorganismal growth in the carriers before and after six cycles, with a uniform distribution pattern and localized agglomeration (partial images are shown in Figure 9), indicating that the carriers protected the microorganisms from damage by the active species and the illuminant radiation. However, the distributions of biofilm in terms of quantity and distance were relatively inferior to those of the correlational studies, and a possible reason for the short one-week time for loading microorganisms was the limitations due to the experimental time. Therefore, in order to enhance the contribution to microflora, adequate time needs to be allowed.

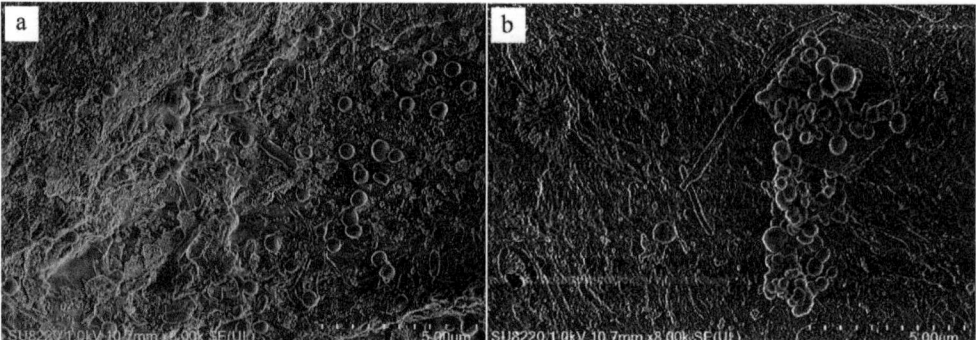

Figure 9. SEM images of biofilms located in carriers before (**a**) and after (**b**) ICPB.

3.5.2. Genera Composition of Microbial Community

The community composition and succession are shown in Figure 10, with the relative abundance of bacteria greater than 1% at the level of genus, which includes the species of samples of the initial community, the community after biodegradation alone (B), and the community after ICPB system. Methyloversatilis genus belongs to the Proteobacteria phylum and its relative abundance decreases dramatically from the starting point to the B and ICPB systems. The results indicate that the B and ICPB systems were not conducive to the growth of the Methyloversatilis genus. Studies have shown that the Methyloversatilis genus is the dominant bacteria in the environment [39–41] with low-carbon organic compounds, such as methylamine, acetone, dichloroethane, etc., and the carbon sourced from the B and ICPB systems was less than that sourced from the initial system. Therefore, it was difficult to meet the metabolic demands of a large number of microorganisms and the relative abundance decreased significantly. Moreover, the Methyloversatilis genus could participate in the reduction of ClO_4^- [42] and play an important role in the microbial degradation of aromatic compounds such as benzene and naphthalene [43,44], organic pesticides [45], and petroleum pollutants [46]. Therefore, the Methyloversatilis genus had a certain capacity for oxidation and chlorine resistance and could survive and contribute to 1,2,4-TrCB degradation in the B and ICPB systems. In both the B and ICPB systems, the growth of the Methyloversatilis genus was inhibited and the Sediminibacterium genus gradually became the dominant bacteria. The Sediminibacterium genus belongs to the Bacteroidota phylum and is a widespread genus in the environment ranging from soils and sediments [47] to lakes and reservoirs [48,49], mining wastewater [50], and urban rivers [51]. Hence, it plays an important role in water purification and the carbon cycle. Moreover, the Sediminibacterium genus was found to be one of the dominant bacteria in the storage pool of radioactive materials [52], and the reclaimed water was treated with a low concentration of chlorine [53], signifying its oxidation resistance. However, the dominant bacteria were replaced by the Acidovorax genus when the reclaimed water was treated with ultraviolet and chlorine together [54], indicating that ultraviolet light had adverse effects on the Sediminibacterium genus. This provides a possible explanation for why the relative abundance of this genus decreased by 13.13% in the ICPB system compared to the B system, which is that it was due to illuminant damage. In addition, the Sediminibacterium genus could use many simple and complex organic compounds, such as vinyl chloride as a carbon source necessary for growth [55,56]. Therefore, the extensive growth potential caused it to dominate in the B and ICPB systems.

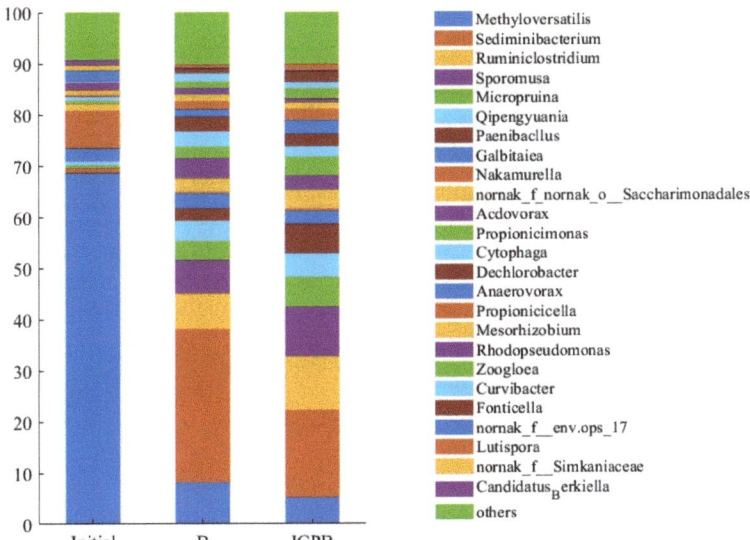

Figure 10. Composition of microbial communities at genus level in samples of initial, B, and ICPB. Initial—the sample before the reaction; B and ICPB—the samples of the biodegradation and coupling, respectively.

The Ruminiclostridium genus, a kind of anaerobes, is commonly found in the anaerobic digestion system and can use crop stalks and other cellulose as a carbon source for anaerobic metabolism [57–59]. The main component of the carriers in this experiment was cellulose, which provided the conditions for the growth of the Ruminiclostridium genus. The internal structure of the carriers may be changed due to the long amount of time needed to run the ICPB system. In addition, considering the carbon source supplied to the system enhancing the metabolic activity of the microbial community, the acetic and propionic acids [60] produced by the Ruminiclostridium genus improved the degradation efficiency of 1,2,4-TrCB. However, it was not clear whether the Ruminiclostridium genus directly degraded the 1,2,4-TrCB or not. The Sporomusa genus, a kind of anaerobic bacteria belonging to Firmicutes that is widely distributed over hypoxic sediments of freshwater rivers, lakes, streams, and ditches, used a variety of electron donors for metabolism. The Sporomusa genus produced Cobamides with the functions of carbon skeleton rearrangement, methyl transfer, and reductive dechlorination [61]. A study has shown that the Sporomusa genus is conducive to the dichlorination of the Dechloromonas genus [62]. Furthermore, while degrading the pollutants, it transformed carbon dioxide into acetic acid and hydrogen, providing electron donors for other microorganisms and making a difference in the microbial community.

Other bacterial genera with a relative abundance of more than 1% are shown in Figure 11. It is evident that microflora succession occurred during the degradation process of 1,2,4-TrCB. Compared with the sample of initial microbial community, a variety of genera that may be involved in the degradation of 1,2,4-TrCB were enriched in the B and ICPB systems, mainly the Micropruina genus, Qipengyuania genus, Paenibacllus genus, and Acidovorax genus. The Micropruina genus had a strong ability for total tolerance and stored lactic acid, acetic acid, and ethanol as glycogen, making a great contribution to the removal of the phosphorus and chemical oxygen demand (COD) from water bodies [63]. The Qipengyuania genus is widely distributed in freshwater, seawater, sediment, and other environments and has functional genes for the cycling of nitrogen, sulfur, and phosphorus [64]. The Qipengyuania genus has the potential to degrade microbial soluble metabolites and there are a few studies on water resource purification. The Paenibacllus genus resisted the negative influence of extreme temperatures, pH, pressure, and ultraviolet radiation due to spore formation for reproduction [65] and participated in the degradation of organic pesticides and polycyclic aromatic hydrocarbons [66]. Acidovorax, which can alleviate the oxidation of free radicals, is commonly seen in studies about the degradation of pollutants and has a significant effect on the biodegradation of aromatic compounds such as phenols, biphenyl, and chlorobenzene [67]. Therefore, the microbial community was involved in the degradation of 1,2,4-TrCB with reasonable speculation and improved the removal rate of 1,2,4-TrCB in the ICPB (Figure 6). Moreover, a functional community structure containing aerobic, facultative, and anaerobic microorganisms was constructed in the ICPB with high stability for pollutant degradation. The community structure of the biofilm in the carriers changed in the ICPB with the dominant microorganisms varying from the Methyloversatilis and Nakamurella genera to the Sediminibacterium, Ruminiclostridium, Sporomusa, and Methyloversatilis genera.

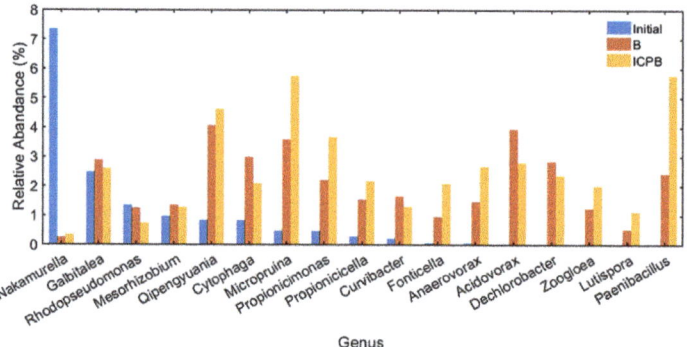

Figure 11. Other genera with a relative abundance of more than 1% in the samples of initial, B, and ICPB. Initial—the sample before the reaction; B and ICPB—the samples of the biodegradation and coupling, respectively.

3.5.3. Correlation Analysis of Microbial Community

The correlation analysis of the microflora is shown in Figure 12. The community heatmap analysis shows not only the kinships of the main genera but also the relative abundances of the main genera of the different samples (initial, B, and ICPB); the color gradient from blue to red corresponds to the relative abundances from low to high. The network plot shows the correlations between the main genera; the green and red lines represent negative and positive correlations, respectively, and the nodes' sizes represent the relative abundances. There was a significant correlation between the top 20 bacteria in the three samples ($p < 0.05$, $R > 0.9$). The average degree of the nodes was 9.16 and the average clustering coefficient was 1, indicating that all of the nodes were connected with the other nodes and that each genus played an important role in the community. Moreover, the microorganisms with close kinships were consistent with each other when their abundances and correlations with the others changed. The Methyloversatilis, Rhodopseudomonas, and Nakamurella genera had the same abundance variation trends for the three samples and were negatively correlated with the other bacteria connected by lines. The competitive relationship with the others could come from the three genera. Therefore, a possible kind of reference to optimize the structure of the microbial community was provided by the consistent changes in the microorganisms.

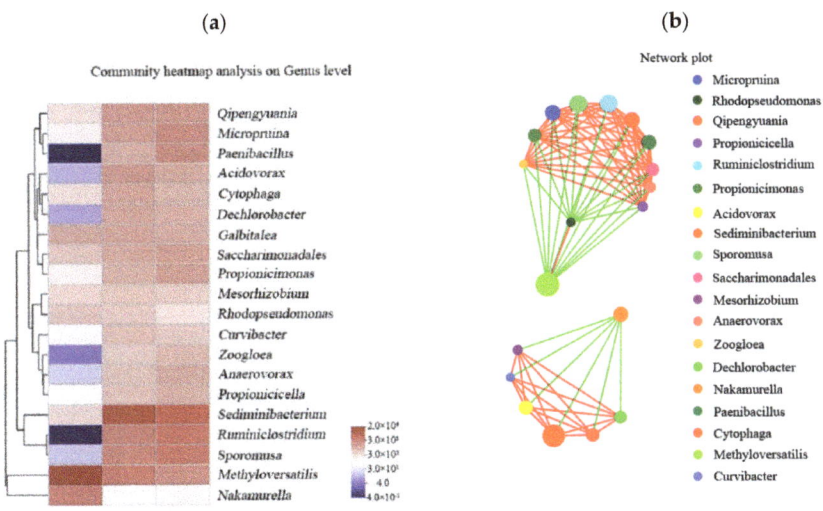

Figure 12. Correlation analysis of microflora. (**a**) Image of heatmap and relative abundances from low to high with colors varying from blue to red; (**b**) Network plot of the community, with each point referring to one kind of genus. The red and blue lines refer to the positive and negative correlations, respectively.

3.6. Degradation Mechanism of 1,2,4-TrCB in ICPB

In conclusion, considering the studies on pyridine [68], trichlorophenol [69], nitrophenol [29], and other pollutants in the ICPB system, a synergistic effect was found between adsorption, photocatalysis, and biodegradation [70]. In the system, 1,2,4-TrCB first generated low chlorobenzene, chlorophenol, and other intermediates with simple structures and low toxicity through reactions. These substances further reacted to generate phenols and esters, and then the benzene ring broke to generate alcohols, fatty acids, etc. Finally, these simple organic substances were mineralized into CO_2 and H_2O. In this process, the porous carriers protected the microorganisms from the adverse effects of photocatalysis; microbial metabolism usually increases opportunities for interactions between the active groups and the substrates. In the system, the h^+ radical contributed the most to the degradation of 1,2,4-TrCB, followed by $\cdot OH$ and $\cdot O_2^-$. In the microbial community, the dominant bacteria were the Metallopolitalis and Rhodopseudomonas genera, which played an important role in the degradation of 1,2,4-TrCB. The ICPB provided the advantages of robust oxidability and absolute mineralization and therefore has important research value and application prospects in the removal of refractory organics.

4. Conclusions

(1) The ICPB system was constructed using a sugarcane cellulose-TiO_2 carrier. This technology played an important role in promoting the degradation of 1,2,4-TrCB. Compared with biodegradation and photocatalysis alone, the removal rates of 1,2,4-TrCB increased by 68.01% and 14.81%, respectively, and the mineralization rates increased by 50.30% and 11.50%, respectively.

(2) The sugarcane cellulose-TiO_2 carrier protected the dominant bacteria in the biofilm from damage by photocatalysis. The microorganism decomposed some of the photocatalysis products, making more free radicals that were used for the degradation of the intermediate products, thus improving the degradation rate and mineralization rate of 1,2,4-TrCB.

Author Contributions: Conceptualization: Z.Z., C.J. and J.X.; methodology: Y.L., Z.Z. and C.J.; formal analysis and investigation: A.D., J.Z. and Z.Z.; writing—original draft preparation: C.J. and Z.Z.; Writing—review and editing: H.Z.; Funding acquisition: G.C. and L.L.; Supervision: J.X. All authors have read and agreed to the published version of the manuscript.

Funding: The authors wish to thank the National Natural Science Foundation of China (NSFC No:21968005), the Innovation Project of Guangxi Graduate Education (YCSW2022055), Guangxi Ba-Gui Scholars Program (2019A33), the Guangxi Major Projects of Science and Technology (Grant No. GXMPSTAB21196064), the foundation of Guangxi Key Laboratory of Clean Pulp & Papermaking and Pollution Control.

Conflicts of Interest: The authors declare that they have no competing interests.

References

1. Shang, X.; Yang, L.; Ouyang, D.; Zhang, B.; Zhang, W.; Gu, M.; Li, J.; Chen, M.; Huang, L.; Qian, L. Enhanced removal of 1,2,4-trichlorobenzene by modified biochar supported nanoscale zero-valent iron and palladium. *Chemosphere* **2020**, *249*, 126518. [CrossRef] [PubMed]
2. Niu, Y.; Liu, L.; Wang, F.; Liu, X.; Huang, Z.; Zhao, H.; Qi, B.; Zhang, G. Exogenous silicon enhances resistance to 1,2,4-trichlorobenzene in rice. *Sci. Total Environ.* **2022**, *845*, 157248. [CrossRef]
3. Conte, L.O.; Dominguez, C.M.; Checa-Fernandez, A.; Santos, A. Vis LED Photo-Fenton Degradation of 124-Trichlorobenzene at a Neutral pH Using Ferrioxalate as Catalyst. *Int. J. Environ. Res. Public Health* **2022**, *19*, 9733. [CrossRef] [PubMed]
4. Egorova, D.O.; Nazarova, E.A.; Demakov, V.A. New Lindane-Degrading Strains *Achromobacter* sp. NE1 and *Brevundimonas* sp. 242. *Microbiology* **2021**, *90*, 392–396. [CrossRef]
5. Liu, Z.; Zhou, N.; Qiao, W.; Ye, S. Effects of o-nitro-p-methylphenol and o-amino-p-methylphenol on the anaerobic biodegradation of 1,2,4-TCB. *Earth Sci. Front.* **2021**, *28*, 159–166.
6. Qian, Y.; Kong, X. Research Progresses in Environmental Remediation Technologies for 1,2,4-Trichlorobenzene Pollution. *Environ. Prot. Chem. Ind.* **2015**, *35*, 147–153.
7. Marsolek, M.D.; Torres, C.I.; Hausner, M.; Rittmann, B.E. Intimate coupling of photocatalysis and biodegradation in a photocatalytic circulating-bed biofilm reactor. *Biotechnol. Bioeng.* **2008**, *101*, 83–92. [CrossRef]
8. Ma, D.; Zou, D.; Zhou, D.; Li, T.; Dong, S.; Xu, Z.; Dong, S. Phenol removal and biofilm response in coupling of visible-light-driven photocatalysis and biodegradation: Effect of hydrothermal treatment temperature. *Int. Biodeterior. Biodegrad.* **2015**, *104*, 178–185. [CrossRef]
9. Xiong, H.; Zou, D.; Zhou, D.; Dong, S.; Wang, J.; Rittmann, B.E. Enhancing degradation and mineralization of tetracycline using intimately coupled photocatalysis and biodegradation (ICPB). *Chem. Eng. J.* **2017**, *316*, 7–14. [CrossRef]
10. Zhang, C.; Fu, L.; Xu, Z.; Xiong, H.; Zhou, D.; Huo, M. Contrasting roles of phenol and pyrocatechol on the degradation of 4-chlorophenol in a photocatalytic-biological reactor. *Environ. Sci. Pollut. Res.* **2017**, *24*, 24725–24731. [CrossRef]
11. Lu, Z.; Xu, Y.; Akbari, M.Z.; Liang, C.; Peng, L. Insight into integration of photocatalytic and microbial wastewater treatment technologies for recalcitrant organic pollutants: From sequential to simultaneous reactions. *Chemosphere* **2022**, *295*, 133952. [CrossRef] [PubMed]
12. Li, F.; Lan, X.; Shi, J.; Wang, L. Loofah sponge as an environment-friendly biocarrier for intimately coupled photocatalysis and biodegradation (ICPB). *J. Water Process Eng.* **2021**, *40*, 101965. [CrossRef]
13. Zhou, D.; Xu, Z.; Dong, S.; Huo, M.; Dong, S.; Tian, X.; Cui, B.; Xiong, H.; Li, T.; Ma, D. Intimate Coupling of Photocatalysis and Biodegradation for Degrading Phenol Using Different Light Types: Visible Light vs. UV Light. *Environ. Sci. Technol.* **2015**, *49*, 7776–7783. [CrossRef] [PubMed]
14. Zhang, L.; Xing, Z.; Zhang, H.; Li, Z.; Wu, X.; Zhang, X.; Zhang, Y.; Zhou, W. High thermostable ordered mesoporous SiO$_2$-TiO$_2$ coated circulating-bed biofilm reactor for unpredictable photocatalytic and biocatalytic performance. *Appl. Catal. B Environ.* **2016**, *180*, 521–529. [CrossRef]
15. Li, Y.; Zhang, C.; Shuai, D.; Naraginti, S.; Wang, D.; Zhang, W. Visible-light-driven photocatalytic inactivation of MS2 by metal-free g-C$_3$N$_4$: Virucidal performance and mechanism. *Water Res.* **2016**, *106*, 249–258. [CrossRef]
16. Wang, C.; Ren, B.; Hursthouse, A.S.; Hou, B.; Peng, Y. Visible Light-Driven Photocatalytic Degradation of 1,2,4-trichlorobenzene with Synthesized Co$_3$O$_4$ Photocatalyst. *Pol. J. Environ. Stud.* **2018**, *27*, 2285–2292. [CrossRef]
17. Kozhevnikova, N.S.; Gorbunova, T.I.; Vorokh, A.S.; Pervova, M.G.; Zapevalov, A.Y.; Saloutin, V.I.; Chupakhin, O.N. Nanocrystalline TiO$_2$ doped by small amount of pre-synthesized colloidal CdS nanoparticles for photocatalytic degradation of 1,2,4-trichlorobenzene. *Sustain. Chem. Pharm.* **2019**, *11*, 1–11. [CrossRef]
18. Zhang, W.; Shi, Z.; Zhang, Q.; Zhang, H.; Xu, S.-K. Photocatalytic degradation of 1,2, 4-trichlorobenzene with TiO2 coated on carbon nanotubes. *Huan Jing Ke Xue = Huanjing Kexue* **2011**, *32*, 1974–1979.
19. Shi, Z.; Zhang, Q.; Zhang, W.; Xu, S.-K.; Zhang, H. Study on photocatalytic degradation of 1, 2, 3-trichlorobenzene using the microwaved MWNTs/TiO2 composite. *Huan Jing Ke Xue = Huanjing Kexue* **2012**, *33*, 3840–3846.
20. Brunsbach, F.R.; Reineke, W. Degradation of chlorobenzenes in soil slurry by a specialized organism. *Appl. Microbiol. Biotechnol.* **1994**, *42*, 415–420. [CrossRef]

21. Dermietzel, J.; Vieth, A. Chloroaromatics in groundwater: Chances of bioremediation. *Environ. Geol.* **2002**, *41*, 683–689.
22. Dong, W.H.; Zhang, P.; Lin, X.Y.; Zhang, Y.; Taboure, A. Natural attenuation of 1,2,4-trichlorobenzene in shallow aquifer at the Luhuagang's landfill site, Kaifeng, China. *Sci. Total Environ.* **2015**, *505*, 216–222. [CrossRef] [PubMed]
23. Mao, W.; Zhang, L.; Wang, T.; Bai, Y.; Guan, Y. Fabrication of highly efficient Bi_2WO_6/CuS composite for visible-light photocatalytic removal of organic pollutants and Cr(VI) from wastewater. *Front. Environ. Sci. Eng.* **2021**, *15*, 52. [CrossRef]
24. Amini Herab, A.; Salari, D.; Tseng, H.-H.; Niaei, A.; Mehrizadeh, H.; Rahimi Aghdam, T. Synthesis of $BiFeO_3$ nanoparticles for the photocatalytic removal of chlorobenzene and a study of the effective parameters. *React. Kinet. Mech. Catal.* **2020**, *131*, 437–452. [CrossRef]
25. Pan, X.; Wei, J.; Qu, R.; Xu, S.; Chen, J.; Al-Basher, G.; Li, C.; Shad, A.; Dar, A.A.; Wang, Z. Alumina-mediated photocatalytic degradation of hexachlorobenzene in aqueous system: Kinetics and mechanism. *Chemosphere* **2020**, *257*, 127256. [CrossRef] [PubMed]
26. Song, L.; Xing, L. Adsorption and Degradation of 1, 2, 4-Trichlorobenzene by Activated Carbon-Microorganisms in Soil. *J. Agro-Environ. Sci.* **2015**, *34*, 1535–1541.
27. Zhang, Y.; Wang, L.; Rittmann, B.E. Integrated photocatalytic-biological reactor for accelerated phenol mineralization. *Appl. Microbiol. Biotechnol.* **2010**, *86*, 1977–1985. [CrossRef]
28. Zhang, Y.; Sun, X.; Chen, L.; Rittmann, B.E. Integrated photocatalytic-biological reactor for accelerated 2,4,6-trichlorophenol degradation and mineralization. *Biodegradation* **2012**, *23*, 189–198. [CrossRef]
29. Wen, D.; Li, G.; Xing, R.; Park, S.; Rittmann, B.E. 2,4-DNT removal in intimately coupled photobiocatalysis: The roles of adsorption, photolysis, photocatalysis, and biotransformation. *Appl. Microbiol. Biotechnol.* **2012**, *95*, 263–272. [CrossRef]
30. Korosi, L.; Bognar, B.; Bouderias, S.; Castelli, A.; Scarpellini, A.; Pasquale, L.; Prato, M. Highly-efficient photocatalytic generation of superoxide radicals by phase-pure rutile TiO_2 nanoparticles for azo dye removal. *Appl. Surf. Sci.* **2019**, *493*, 719–728. [CrossRef]
31. Liu, X.; Kong, L.; Hujiabudula, M.; Xu, S.; Ma, F.; Zhong, M.; Liu, J.; Abulikemu, A. Progress in preparation and photocatalytic properties of molybdate. *Appl. Chem. Ind.* **2021**, *50*, 217–224.
32. Sobczynski, A.; Dobosz, A. Water purification by photocatalysis on semiconductors. *Pol. J. Environ. Stud.* **2001**, *10*, 195–205.
33. Luan, Y.; Jing, L.; Meng, Q.; Nan, H.; Luan, P.; Xie, M.; Feng, Y. Synthesis of Efficient Nanosized Rutile TiO_2 and Its Main Factors Determining Its Photodegradation Activity: Roles of Residual Chloride and Adsorbed Oxygen. *J. Phys. Chem. C* **2012**, *116*, 17094–17100. [CrossRef]
34. Lichtenberger, J.; Amiridis, M.D. Catalytic oxidation of chlorinated benzenes over V_2O_5/TiO_2 catalysts. *J. Catal.* **2004**, *223*, 296–308. [CrossRef]
35. Lin, S.; Su, G.; Zheng, M.; Jia, M.; Qi, C.; Li, W. The degradation of 1,2,4-trichlorobenzene using synthesized Co_3O_4 and the hypothesized mechanism. *J. Hazard. Mater.* **2011**, *192*, 1697–1704. [CrossRef] [PubMed]
36. Zhu, T.; Zhang, X.; Ma, M.; Wang, L.; Xue, Z.; Hou, Y.; Ye, Z.; Liu, T. 1,2,4-Trichlorobenzene decomposition using non-thermal plasma technology. *Plasma Sci. Technol.* **2020**, *22*, 034011. [CrossRef]
37. Feng, Q.; Wang, Q.; Yu, Y.; Jing, Z. Catalytic Degradation of 1,2,4-Trichlorobenzene with Co-Mn Polyhedral Composite Oxides by Sol-Gel Process. *Chem. World* **2017**, *58*, 7–12.
38. Zhang, W.; Li, L.; Zhang, Q.; Xu, S.; Zhang, H. Study on photocatalytic degradation of typical chlorobenzenes with $MWNTs/TiO_2$. *Acta Sci. Circumstantiae* **2012**, *32*, 631–638.
39. Doronina, N.V.; Kaparullina, E.N.; Trotsenko, Y.A. Methyloversatilis thermotolerans sp. nov. a novel thermotolerant facultative methylotroph isolated from a hot spring. *Int. J. Syst. Evol. Microbiol.* **2014**, *64*, 158–164. [CrossRef]
40. Lu, H.; Kalyuzhnaya, M.; Chandran, K. Comparative proteomic analysis reveals insights into anoxic growth of Methyloversatilis universalis FAM5 on methanol and ethanol. *Environ. Microbiol.* **2012**, *14*, 2935–2945. [CrossRef]
41. Smalley, N.E.; Taipale, S.; De Marco, P.; Doronina, N.V.; Kyrpides, N.; Shapiro, N.; Woyke, T.; Kalyuzhnaya, M.G. Functional and genomic diversity of methylotrophic Rhodocyclaceae: Description of Methyloversatilis discipulorum sp nov. *Int. J. Syst. Evol. Microbiol.* **2015**, *65*, 2227–2233. [CrossRef] [PubMed]
42. Li, H.; Zhou, L.; Lin, H.; Zhang, W.; Xia, S. Nitrate effects on perchlorate reduction in a H_2/CO_2-based biofilm. *Sci. Total Environ.* **2019**, *694*, 133564. [CrossRef] [PubMed]
43. Kittichotirat, W.; Good, N.M.; Hall, R.; Bringel, F.; Lajus, A.; Medigue, C.; Smalley, N.E.; Beck, D.; Bumgarner, R.; Vuilleumier, S.; et al. Genome Sequence of Methyloversatilis universalis FAM5^T, a Methylotrophic Representative of the Order Rhodocyclales. *J. Bacteriol.* **2011**, *193*, 4541–4542. [CrossRef] [PubMed]
44. Rochman, F.F.; Sheremet, A.; Tamas, I.; Saidi-Mehrabad, A.; Kim, J.-J.; Dong, X.; Sensen, C.W.; Gieg, L.M.; Dunfield, P.F. Benzene and Naphthalene Degrading Bacterial Communities in an Oil Sands Tailings Pond. *Front. Microbiol.* **2017**, *8*, 1845. [CrossRef] [PubMed]
45. Qi, Z.; Wei, Z. Microbial flora analysis for the degradation of beta-cypermethrin. *Environ. Sci. Pollut. Res.* **2017**, *24*, 6554–6562. [CrossRef]
46. Jiao, S.; Liu, Z.; Lin, Y.; Yang, J.; Chen, W.; Wei, G. Bacterial communities in oil contaminated soils: Biogeography and co-occurrence patterns. *Soil Biol. Biochem.* **2016**, *98*, 64–73. [CrossRef]
47. Wu, Y.; Lin, H.; Yin, W.; Shao, S.; Lv, S.; Hu, Y. Water Quality and Microbial Community Changes in an Urban River after Micro-Nano Bubble Technology in Situ Treatment. *Water* **2019**, *11*, 66. [CrossRef]

48. Guo, D.; Liang, J.; Chen, W.; Wang, J.; Ji, B.; Luo, S. Bacterial Community Analysis of Two Neighboring Freshwater Lakes Originating from One Lake. *Pol. J. Environ. Stud.* **2021**, *30*, 111–117. [CrossRef]
49. Kang, H.; Kim, H.; Lee, B.-I.; Joung, Y.; Joh, K. *Sediminibacterium goheungense* sp nov. isolated from a freshwater reservoir. *Int. J. Syst. Evol. Microbiol.* **2014**, *64*, 1328–1333. [CrossRef]
50. Ettamimi, S.; Carlier, J.D.; Cox, C.J.; Elamine, Y.; Hammani, K.; Ghazal, H.; Costa, M.C. A meta-taxonomic investigation of the prokaryotic diversity of water bodies impacted by acid mine drainage from the SAo Domingos mine in southern Portugal. *Extremophiles* **2019**, *23*, 821–834. [CrossRef]
51. Reza, M.S.; Mizusawa, N.; Kumano, A.; Oikawa, C.; Ouchi, D.; Kobiyama, A.; Yamada, Y.; Ikeda, Y.; Ikeda, D.; Ikeo, K.; et al. Metagenomic analysis using 16S ribosomal RNA genes of a bacterial community in an urban stream, the Tama River, Tokyo. *Fish. Sci.* **2018**, *84*, 563–577. [CrossRef]
52. Ruiz-Lopez, S.; Foster, L.; Boothman, C.; Cole, N.; Morris, K.; Lloyd, J.R. Identification of a Stable Hydrogen-Driven Microbiome in a Highly Radioactive Storage Facility on the Sellafield Site. *Front. Microbiol.* **2020**, *11*, 587556. [CrossRef] [PubMed]
53. Li, Y.; Wang, F.; Yang, H.; Pan, M.; Du, J. Study on Microbial Community Composition and Variation based on High Throughput Sequencing under Leersia hexandra Swartz Ecological Floating Bed. *Southwest China J. Agric. Sci.* **2018**, *31*, 1903–1911.
54. Wang, H.; Hu, C.; Hu, X. Effects of combined UV and chlorine disinfection on corrosion and water quality within reclaimed water distribution systems. *Eng. Fail. Anal.* **2014**, *39*, 12–20. [CrossRef]
55. Singleton, D.R.; Adrion, A.C.; Aitken, M.D. Surfactant-induced bacterial community changes correlated with increased polycyclic aromatic hydrocarbon degradation in contaminated soil. *Appl. Microbiol. Biotechnol.* **2016**, *100*, 10165–10177. [CrossRef] [PubMed]
56. Wilson, F.P.; Liu, X.; Mattes, T.E.; Cupples, A.M. Nocardioides, Sediminibacterium, Aquabacterium, Variovorax, and Pseudomonas linked to carbon uptake during aerobic vinyl chloride biodegradation. *Environ. Sci. Pollut. Res.* **2016**, *23*, 19062–19070. [CrossRef] [PubMed]
57. Cao, G.; Song, T.; Shen, Y.; Jin, Q.; Feng, W.; Fan, L.; Cai, W. Diversity of Bacterial and Fungal Communities in Wheat Straw Compost for Agaricus bisporus Cultivation. *Hortscience* **2019**, *54*, 100–109. [CrossRef]
58. Vita, N.; Ravachol, J.; Franche, N.; Borne, R.; Tardif, C.; Pages, S.; Fierobe, H.-P. Restoration of cellulase activity in the inactive cellulosomal protein Cel9V from Ruminiclostridium cellulolyticum. *Febs Lett.* **2018**, *592*, 190–198. [CrossRef]
59. Zhu, Q.; Dai, L.; Wang, Y.; Tan, F.; Chen, C.; He, M.; Maeda, T. Enrichment of waste sewage sludge for enhancing methane production from cellulose. *Bioresour. Technol.* **2021**, *321*, 124497. [CrossRef]
60. Ziganshina, E.E.; Belostotskiy, D.E.; Bulynina, S.S.; Ziganshin, A.M. Effect of magnetite on anaerobic digestion of distillers grains and beet pulp: Operation of reactors and microbial community dynamics. *J. Biosci. Bioeng.* **2021**, *131*, 290–298. [CrossRef]
61. Newmister, S.A.; Chan, C.H.; Escalante-Semerena, J.C.; Rayment, I. Structural Insights into the Function of the Nicotinate Mononucleotide:phenol/p-cresol Phosphoribosyltransferase (ArsAB) Enzyme from Sporomusa ovata. *Biochemistry* **2012**, *51*, 8571–8582. [CrossRef] [PubMed]
62. Yang, Y.R.; Pesaro, M.; Sigler, W.; Zeyer, J. Identification of microorganisms involved in reductive dehalogenation of chlorinated ethenes in an anaerobic microbial community. *Water Res.* **2005**, *39*, 3954–3966. [CrossRef] [PubMed]
63. McIlroy, S.J.; Onetto, C.A.; McIlroy, B.; Herbst, F.-A.; Dueholm, M.S.; Kirkegaard, R.H.; Fernando, E.; Karst, S.M.; Nierychlo, M.; Kristensen, J.M.; et al. Genomic and in Situ Analyses Reveal the *Micropruina* spp. as Abundant Fermentative Glycogen Accumulating Organisms in Enhanced Biological Phosphorus Removal Systems. *Front. Microbiol.* **2018**, *9*, 1004. [CrossRef] [PubMed]
64. Feng, X.-M.; Mo, Y.-X.; Han, L.; Nogi, Y.; Zhu, Y.-H.; Lv, J. Qipengyuania sediminis gen. nov. sp nov. a member of the family Erythrobacteraceae isolated from subterrestrial sediment. *Int. J. Syst. Evol. Microbiol.* **2015**, *65*, 3658–3665. [CrossRef]
65. Amoah, K.; Huang, Q.-c.; Dong, X.-h.; Tan, B.-p.; Zhang, S.; Chi, S.-y.; Yang, Q.-h.; Liu, H.-y.; Yang, Y.-z. Paenibacillus polymyxa improves the growth, immune and antioxidant activity, intestinal health, and disease resistance in Litopenaeus vannamei challenged with Vibrio parahaemolyticus. *Aquaculture* **2020**, *518*, 734563. [CrossRef]
66. Nwinyi, O.C.; Amund, O.O. Biodegradation of Selected Polycyclic Aromatic Hydrocarbons by Axenic Bacterial Species Belonging to the Genera Lysinibacillus and Paenibacillus. *Iran. J. Sci. Technol. Trans. A Sci.* **2017**, *41*, 577–587. [CrossRef]
67. Shehu, D.; Alias, Z. Dechlorination of polychlorobiphenyl degradation metabolites by a recombinant glutathione S-transferase from Acidovorax sp. KKS102. *Febs Open Bio* **2019**, *9*, 408–419. [CrossRef]
68. Fang, M.-M.; Yan, N.; Zhang, Y.-M. Biodegradation of pyridine under UV irradiation. *Huan Jing Ke Xue = Huanjing Kexue* **2012**, *33*, 488–494.
69. Li, G.; Park, S.; Kang, D.-W.; Krajmalnik-Brown, R.; Rittmann, B.E. 2,4,5-Trichlorophenol Degradation Using a Novel TiO2-Coated Biofilm Carrier: Roles of Adsorption, Photocatalysis, and Biodegradation. *Environ. Sci. Technol.* **2011**, *45*, 8359–8367. [CrossRef]
70. Yu, M.; Wang, J.; Tang, L.; Feng, C.; Liu, H.; Zhang, H.; Peng, B.; Chen, Z.; Xie, Q. Intimate coupling of photocatalysis and biodegradation for wastewater treatment: Mechanisms, recent advances and environmental applications. *Water Res.* **2020**, *175*, 115673. [CrossRef]

Article

Accurate Determination of Moisture Content in Flavor Microcapsules Using Headspace Gas Chromatography

Xueyan Liu [1,†], Chuxing Zhu [2,†], Kang Yu [1], Wei Li [1], Yingchun Luo [1], Yi Dai [1,2,*] and Hao Wang [3,*]

[1] School of Chemical Engineering, Guizhou Minzu University, Guiyang 550025, China; liu1996@gzmu.edu.cn (X.L.); yukang@gzmu.edu.cn (K.Y.); liwei1998@gzmu.edu.cn (W.L.); 05111583@gzmu.edu.cn (Y.L.)
[2] State Key Laboratory of Pulp and Paper Engineering, South China University of Technology, Guangzhou 510641, China; 202010105766@mail.scut.edu.cn
[3] Technology Center, China Tobacco Yunnan Industrial Co., Ltd., Kunming 650231, China
* Correspondence: author. daiyi2020@gzmu.edu.cn (Y.D.); wanghao@ynzy-tobacco.com (H.W.); Tel.: +86-15918412162 (Y.D.); +86-15288221204 (H.W.)
† These authors contributed equally to this work.

Abstract: This study demonstrates an accurate method for determining the moisture content in flavor microcapsules using headspace gas chromatography. The method involves measuring the gas chromatography signals of water from vapor in a headspace vial containing flavor microcapsules at a temperature of 125 °C. The measurements were recorded over four headspace extractions, from which the moisture content in the microcapsule samples was extrapolated via simple vapor-phase calibration. The results revealed that the proposed method demonstrated good precision (a relative standard deviation of <3.11%) and accuracy. The proposed method is accurate, highly sensitive, automated, and suitable for testing the moisture content of flavor microcapsules and related products.

Keywords: tracer water; flavor microcapsule; headspace; gas chromatography

1. Introduction

Microcapsules, with sizes ranging between 1 and 1000 μm, are composed of a core (internal part) and shell (external part) [1,2]. The physical properties of the microcapsule shell effectively protect and control the release of sensitive core materials under a variety of conditions. Therefore, they have wide applications in several fields, such as medicine, agriculture, chemical industries, pharmaceuticals, and food engineering [3–8]. To prepare microcapsules, the materials and processing conditions are important for product quality [9]. Additionally, the amount of water in microcapsules, regarded as an impurity, is critical to their quality and performance [10]. For example, if the content of water in flavor microcapsules is extremely high, the viscosity of the shell materials increases, affecting storage and flavor release in related products [11–13]. Therefore, a method that can effectively determine the moisture content of flavor microcapsules is crucial for their production and quality control.

Conventionally, the water content in flavor microcapsules is determined using the direct oven-drying method [14], in which the water in the flavor microcapsule is removed through evaporation at approximately 105 °C for 3 h in an oven. The amount of water lost can be obtained by weighing the sample before and after drying, thereby allowing the calculation of the moisture content in the sample. However, because certain volatile compounds, such as ethyl acetate and ethanol, are also present in microcapsules, they may be removed during the drying process and can cause significant errors in water content measurements [15]. This issue can be addressed using the Karl Fischer titration method [16], in which a reaction occurs between the titration reagent (SO_2 or I_2 in CH_3OH and pyridine or imidazole buffer medium) and water in the sample. This method demonstrates better reproducibility and is sensitive to the moisture content in the sample compared to the oven-drying method. However, in addition to the requirements of a complex titrator, expensive reagents, and time-consuming procedures complicating this method, certain compounds containing carbonyl groups (commonly present in flavorings) in the matrices can react with the Karl Fisher reagent, affecting measurement accuracy [17]. The low-temperature dry nitrogen purge method has also been applied to determine the water content of microcapsules [18]. In this method, a flavor microcapsule sample was placed in a dish in a test chamber (d = 35 cm, H = 6 cm) maintained under a controlled atmosphere (temperature = 40 °C, relative humidity = 1%), and the water in the sample was eliminated by injecting heated absolute dry nitrogen. By recording the weight at each hour, until the ratio of the initial mass change is less than 0.04%, the water content of the sample can be calculated. This method is accurate; however, it takes longer than 16 h to test one sample, which significantly affects its detection efficiency. Gas chromatography (GC) with a thermal conductivity detector (TCD) was also proposed for determining the water content of microcapsule samples [19]. Herein, the water peak was well separated via GC; therefore, the water content could be quantified. However, before GC measurements can be performed, complicated and time-consuming sample pretreatment procedures, such as solvent extraction and separation, are mandatory. This involves the extraction of the sample with isopropanol and filtering of the supernatant liquid using an organic phase filter membrane before quantification using GC.

Headspace gas chromatography (HS-GC), which is different from the conventional GC technique, is an effective tool to detect volatile analytes in solid or liquid complex matrices [20,21]. When equipped with a TCD, HS-GC can be used to determine the water content of samples. However, traditional HS-GC analysis cannot be used to determine the water content in solid samples because the vapor–solid partitioning (K) of water varies with different sample matrices, which introduces errors in the method calibration. Multiple headspace extraction (MHE) modes are available in several commercial headspace auto-samplers [22–24]. For each headspace extraction, the analyte in the vapor phase was partially removed from the vial for GC analysis. If the analyte in the matrices is almost completely vaporized in the vial, the total amount of analyte can be calculated by integrating the contents during the MHE. Consequently, the sample–matrix effect can be eliminated. This suggests that the MHE-GC technique can be used to determine the moisture content of flavor microcapsules.

The objective of this study was to develop an accurate method for the determination of moisture content in flavor microcapsule samples using MHE-GC. The major focus areas involved the establishment of a methodology and determination of the effects of GC conditions, equilibration time, temperature, sample size, and extraction number on the measurement of moisture content. To evaluate the accuracy of this method, microcapsule samples with different flavors were analyzed using MHE-GC, and the results were compared with those obtained using reference methods.

2. Experimental Section

2.1. Materials

Flavor microcapsule samples were collected from EnDian Science and Technology Development of Yunnan Co., Ltd. (Kunming, China). The polymer shell materials were prepared using carboxymethyl chitosan and sodium alginate. The core materials were composed of flavor components and octyl and decyl glycerates. Distilled water was obtained from a laboratory device. All flavor microcapsule samples were stored in sealed bags prior to the analysis.

2.2. Equipment and Operation Procedures

The HS-GC measurements were performed using an automated headspace sampler (DANI HS 86.50 PLUS, Cologno Monzese, Italy) and GC system (Agilent GC 8860A, Santa Clara, CA, USA) equipped with a TCD and capillary column (Model GS-Q, 30 m length × 0.3 μm i.d,

530 µm thickness), J&W Scientific, Folsom, CA, USA). The volumes of the headspace vials and sample loop were 21.6 and 3 mL, respectively. The MHE-GC measurement conditions were as follows: the temperatures of the oven, sample loop, and transportation line were 125, 130, and 135 °C, respectively; both the injector and detector temperatures were 250 °C; the GC injection mode was split, and the split ratio was 10:1; the carrier gas was nitrogen with a flow rate of 25 mL/min; the vial pressurization time was 0.2 min; the loop equilibration time was 0.05 min, and the sample-loop time was 0.2 min; the carrier gas pressure and pressurization were 1.5 and 2.00 bar, respectively; the headspace vial was strongly shaken during equilibration.

2.3. Measurement Procedures

Approximately 0.20 g of the microcapsule sample was added to a dried and empty headspace vial, as presented in Figure 1. The weight of the microcapsule sample was determined by weighing the sample vial (including the PTFE/silicone septum and aluminum cap) before and after sample addition. Subsequently, the sample vial was immediately sealed and transferred to a headspace autosampler for MHE-GC testing. The sample vials were equilibrated at 125 °C, and the interval time for MHE-GC measurements was 6 min. Accordingly, the GC signals of the water vapor (peak area) were measured using GC-TCD for each headspace extraction.

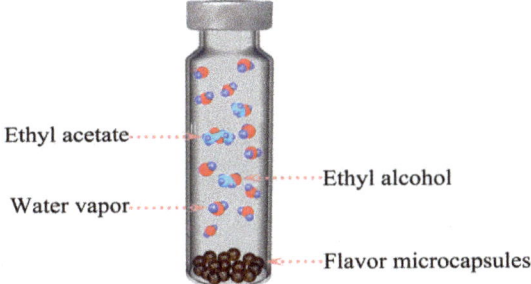

Figure 1. Schematic of the headspace vial used to measure the moisture content in microcapsule samples.

2.4. Determination of Moisture Content Using the Low Temperature Dry Nitrogen Purge Method

The water content in the microcapsule sample was determined using the low-temperature dry nitrogen purge method. The procedure was as follows: the temperature and relative humidity in the chamber were (40 ± 0.1) °C and below 1.0%, respectively. Approximately 5 g of the sample was added to a dish, and the water in the sample was eliminated by injecting heated absolute dry nitrogen. The weight was recorded every hour until the ratio of the mass change to the initial sample mass was less than 0.04%. The water content in the microcapsules (R) was calculated using the following Equation (1):

$$R = \frac{m_0 - m_1}{m_0} \times 100\% \tag{1}$$

where m_0 denotes the weight of the microcapsule sample added to the dish, and m_1 denotes the weight of the sample after reaching a constant weight.

2.5. Determination of Moisture Content Using the Traditional Oven Drying Method

The moisture content of the microcapsule samples was also determined using the oven-drying method. In this method, the test was conducted as follows: approximately 5 g of the sample was added to a dish and placed in a drying oven at 105 °C for approximately 4 h. The dish was then placed in a dryer and allowed to cool to room temperature. This procedure was repeated until a constant weight was obtained. The moisture content in the microcapsules (R) was calculated using the following Equation (2):

$$R = \frac{m_0 - m_1 + m_2}{m_0} \times 100\% \quad (2)$$

where m_0, m_1, and m_2 denote the weights of the microcapsule sample, dish after reaching a constant weight, and the total weight of the dish and microcapsule sample after reaching a constant weight, respectively.

3. Results and Discussion

3.1. Theory of the MHE-GC Method

When a given amount of the flavor microcapsule sample in the sealed headspace vial reached phase equilibrium at an elevated temperature (>100 °C), water co-existed in the solid and gas phases. Using the MHE-GC technique, consecutive analyses were conducted on the same sample vial, from which a part of the water in the gas phase could be extracted. As illustrated in Figure 2, the GC signal intensity of water gradually decreases with more extractions [25–27], which indicates that the entire water content in the microcapsule sample can be completely withdrawn if the extraction number (n) is sufficiently high.

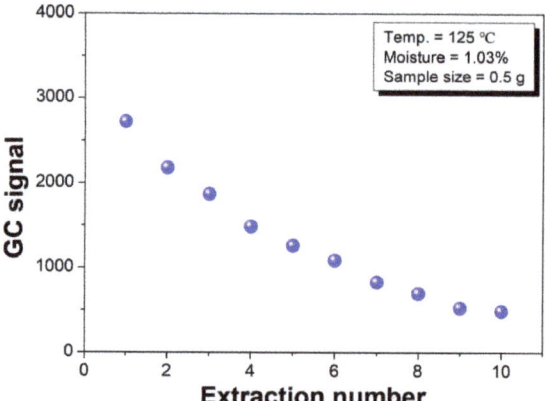

Figure 2. Relationship between the GC signal of water and extraction number.

According to MHE-GC theory [27], the relationship between A_n and n can be described as Equation (3):

$$\mathrm{Log}(A_n) = \mathrm{Log}(A_0) - bn \quad (3)$$

where A_n and A_0 denote the GC signals at n and 0 extractions, respectively, and b denotes the slope obtained from the linear fit. Therefore, the integrated GC signal (A_t) can be expressed as Equation (4):

$$A_t = A_1 + A_1 10^{-b} + \ldots + A_1 10^{-b(n-1)} = \lim_{n \to \infty} \frac{A_1(1 - 10^{-bn})}{1 - 10^{-b}} = \frac{A_1}{1 - 10^{-b}} \quad (4)$$

where A_t denotes the sum of the peak areas, and A_1 indicates the GC signal of the first extraction.

The total mass of water in the gas phase (m_t) is linearly proportional to A_t in Equation (5):

$$m_t = K A_t \quad (5)$$

where K can be obtained using the calibration method.

Therefore, based on Equations (4) and (5), the tracer water content in the microcapsule sample can be calculated as Equation (6):

$$R = \frac{K A_1}{W(1 - 10^{-b})} \times 100\% \quad (6)$$

where R denotes the content of tracer water in the microcapsule sample, and W denotes the weight of the sample added to the headspace vial.

3.2. Selection of HS-GC Measurement Conditions

3.2.1. Conditions for GC Separation

Effective GC separation for water and other volatile substances is required to determine the water content of the microcapsule samples. Figure 3 shows the chromatogram of a microcapsule sample. The water and oxygen (in air) peaks could be separated well under the given GC operation conditions, and no other volatile organic compounds were observed in the chromatogram. These results further support that the GC-TCD system can effectively prevent interference resulting from complex volatile organic compounds in the matrix while determining the amount of water in the samples.

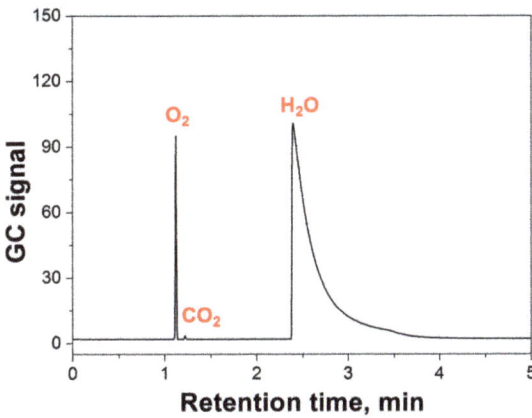

Figure 3. Chromatogram of a microcapsule sample.

3.2.2. Temperature Conditions and Interval Times in Headspace Equilibration

The MHE is based on the gradual removal of water from microcapsules. Therefore, the equilibrium temperature should be higher than the boiling point of water (>100 °C). However, if an extremely high equilibration temperature is applied during the test, it may lead to a high pressure in the sealed headspace vial and increase the risk of leakage [28,29]. Therefore, the feasible equilibration temperature was deemed to be less than 130 °C. Figure 4 depicts the effect of equilibrium temperature on the water lost from a given microcapsule sample during the MHE process, in which two temperatures, 105 °C and 125 °C, were selected. The high temperature (125 °C) accelerated the water removal rate during the sample shaking process, as revealed in other studies [22,25]. Therefore, 125 °C was selected as the equilibration temperature for the subsequent study.

Because the present method is based on the equilibrium of water partitioning between the vapor and solid phases, the time required for equilibration before each headspace extraction must be determined. Figure 5 shows the GC signals of water detected at different time intervals in the headspace vial at a temperature of 125 °C. Equilibrium was attained in 6 min for the microcapsule samples, in which two sample particle sizes of 1 and 3.4 mm were selected. Notably, this equilibrium time is longer than that required for the determination of moisture content in paper materials [22]. This may be related to the structure of the microcapsule shell, which effectively protects the tracer water from the microcapsule release. Therefore, 6 min was chosen as the time interval for the MHE process.

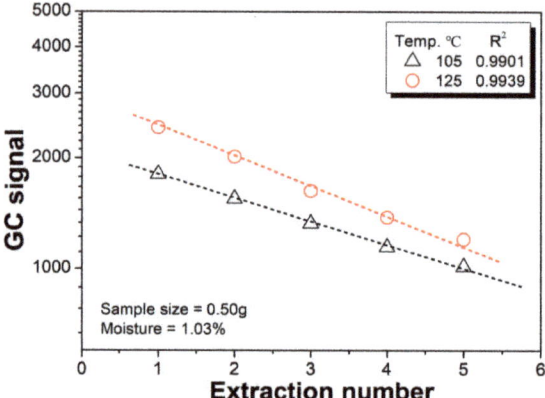

Figure 4. Effect of equilibrium temperature on the water removal.

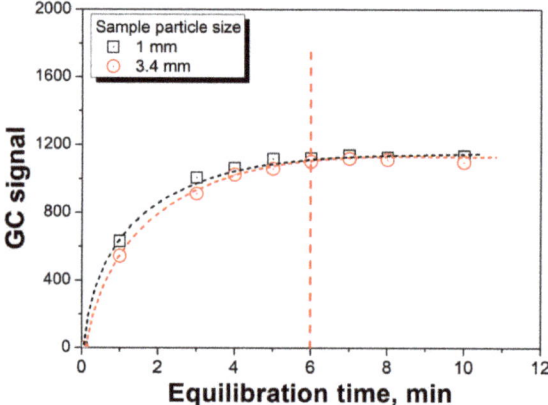

Figure 5. Time required to reach the water headspace equilibrium.

3.2.3. Selection of the Sample Size for Headspace Measurements

In MHE-GC measurements, a large sample weight is helpful for improving the sensitivity of the method [30]. However, using a larger sample size (weight) can cause lower detection efficiency because of the slower release of water from the core material in the flavor microcapsule sample [22]. As mentioned in Section 3.1, the value of A_1 is crucial for calculating the water content in the microcapsule samples. Figure 6 illustrates the effect of the sample size on the water GC signal. A linear variation in the GC signal with respect to the sample size was observed from approximately 0 to 0.51 g. When the sample amount was more than 0.51 g, the A_1 value was lower than the actual value, resulting in an error in the test results. This outcome is because the total gas pressure in the headspace bottle is very high, resulting in pressure effects [30,31]. Therefore, 0.51 g was determined to be the maximum sample content for successfully measuring the moisture content in the samples.

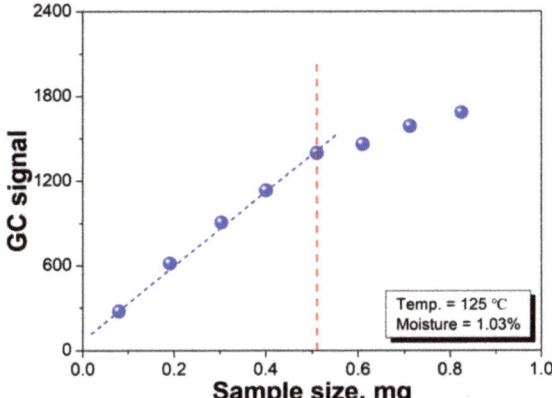

Figure 6. Effect of the sample size on the water GC signal.

3.2.4. Selection of the Extraction Number in the MHE Process

As mentioned in Section 3.1, the value of b in Equation (4) is important for calculating the water content of the microcapsule samples. Figure 7 illustrates the logarithm of the water GC signal versus the extraction number. Consistent with previous studies [32], we obtained a linear relationship between the logarithm of the water GC signal and extraction number. The b value from the linear fit of the 1st to 10th data points is acceptable. The accuracy of the present method can be improved by increasing the extraction number; however, a greater number of extractions leads to longer measurement times, which results in a lower detection efficiency and an increased risk of air leakage. As a compromise, four headspace extractions were selected in the present study.

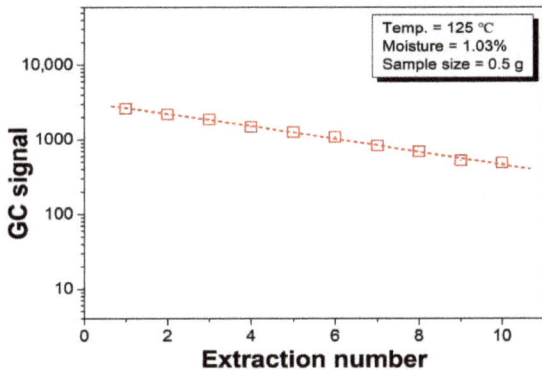

Figure 7. Relationship between the logarithm of the water GC signal and extraction number.

3.3. Method Evaluation

3.3.1. Method Calibration

The HS-GC method was calibrated using a vapor phase calibration technique [22]. Calibration was conducted by adding 0–10.0 mg of distilled water into a set of headspace vials. These vials were tested using the MHE-GC method. Subsequently, the standard external vapor calibration relationship was obtained as follows:

$$A = 190(\pm 2.27) + 2003(\pm 9.01) \times m$$
$$(n = 7, R^2 = 0.9989)$$

(7)

In other words, $A = a(\Delta a) + s(\Delta s) \times m$, where A represents the integrated GC peak areas of water in the MHE-GC measurements, and m denotes the mass (mg) of distilled water added to the headspace vial. s, a, and Δa represent the slope, intercept, and uncertainty, respectively, of the intercept of Equation (7).

The limit of quantitation (LOQ) of the method was 0.0461 mg, which was calculated using Equation (8) [33]. If 0.2 g of microcapsules were used in the testing, the LOQ was approximately 0.0231%. Clearly, the MHE-GC method is highly sensitive and meets the requirements for moisture content testing in the production of flavor microcapsules and related products.

$$LOQ = \frac{a + 10|\Delta a|}{s} \qquad (8)$$

3.3.2. Reproducibility of the Measurements

The reproducibility of the MHE-GC method was investigated by testing three different microcapsule samples in triplicates. As presented in Table 1, the relative standard deviation (RSD) of the water content measured using the proposed method was less than 3.11%, which indicates that the method demonstrates good precision.

Table 1. Reproducibility of the MHE-GC method.

Replica No.	Water Content, %		
	Sample 1	Sample 2	Sample 3
1	1.50	1.13	1.74
2	1.57	1.2	1.71
3	1.56	1.15	1.68
RSD, %	2.45	3.11	1.75

3.3.3. Method Validation

To verify the performance of the present MHE-GC method, seven different particle sizes and types of flavor microcapsules were analyzed using the traditional oven-drying method [14], low-temperature dry nitrogen purge method [18], and MHE-GC method. As presented in Table 2, the results obtained using the low-temperature dry nitrogen purge method matched well with those obtained via the MHE-GC method (relative difference < 5.5%). Importantly, the difference in results is considered minor and significantly below the acceptable values of method errors (e.g., 15%), as revealed through other methods [34,35]. Meanwhile, the values obtained using the MHE-GC method are lower than those obtained through the traditional oven-drying method, which is likely caused by the mass loss of sensitive volatile compounds at high temperatures. Therefore, the traditional oven-drying method is not reliable for determining the moisture content of flavor microcapsules. Conversely, the MHE-GC method is accurate, efficient, and justifiable for the determination of water content in flavor microcapsule samples.

Table 2. Comparison of methods.

Sample ID	Water Content, %			Relative Error, %	
	Low Temperature Dry Nitrogen Purge Method (n = 3)	Oven Drying Method (n = 3)	MHE-GC (n = 3)	Low Temperature Dry Nitrogen Purge Method	Oven Drying Method
1	0.86 ± 0.01	1.01 ± 0.02	0.87 ± 0.02	1.16	−13.9
2	1.12 ± 0.01	1.32 ± 0.01	1.08 ± 0.01	−3.57	−18.2
3	1.40 ± 0.02	1.66 ± 0.03	1.35 ± 0.02	−3.57	−18.7
4	1.66 ± 0.02	1.93 ± 0.02	1.64 ± 0.02	−1.20	−15.0
5	1.99 ± 0.03	2.14 ± 0.03	2.06 ± 0.02	3.52	−3.7
6	1.09 ± 0.03	1.38 ± 0.02	1.03 ± 0.01	−5.50	−25.4
7	2.13 ± 0.01	2.36 ± 0.03	2.05 ± 0.01	−3.76	−13.1

4. Conclusions

An accurate method for determining the moisture content of flavor microcapsule samples was introduced based on the multiple headspace extraction gas chromatography technique. Compared with the traditional oven drying method, the significant advantage of the MHE-GC method is that it is extremely accurate, as it can effectively eliminate the interference caused by the loss of volatile organic compounds in the sample at high temperatures. The results revealed that the proposed method has good precision (a relative standard deviation of <3.11%) and accuracy. The present method is simple and automated and can serve as a reliable tool for testing the moisture content of flavor microcapsule samples and related products.

Author Contributions: X.L.: Writing-Original Draft, Investigation, Methodology, Formal analysis, Validation; C.Z.: Writing-Original Draft, Investigation, Methodology, Formal analysis, Validation; K.Y.: Investigation, Data Curation, Software, Formal analysis; W.L.: Data Curation, Formal analysis, Validation; Y.L.: Visualization, Investigation; Y.D.: Visualization, Supervision, Writing-Review and Editing, H.W.: Investigation and supervision. All authors have read and agreed to the published version of the manuscript.

Funding: This research was funded by Natural Science Foundation of Guizhou Education Commission (Project No. 2022187) and the Natural Science Foundation of Guizhou Minzu University (No. GZMZ2019YB06).

Institutional Review Board Statement: Not applicable.

Informed Consent Statement: Not applicable.

Data Availability Statement: The raw/processed data required to reproduce these findings cannot be shared at this time, as the data also form part of an ongoing study.

Conflicts of Interest: The authors declare that they have no known competing financial interests or personal relationships that could have influenced the work reported in this study.

References

1. Han, R.; Wang, X.; Zhu, G.; Han, N.; Xing, F. Investigation on viscoelastic properties of urea-formaldehyde microcapsules by using nanoindentation. *Polym. Test.* **2019**, *80*, 106146–106156. [CrossRef]
2. Baranauskienė, R.; Bylaitė, E.; Žukauskaitė, J.; Venskutonis, R.P. Flavor Retention of Peppermint (*Mentha piperita* L.) Essential Oil Spray-Dried in Modified Starches during Encapsulation and Storage. *J. Agric. Food Chem.* **2007**, *55*, 3027–3036. [CrossRef] [PubMed]
3. Wang, J.; Yue, X.; Zhang, Z.; Yang, Z.; Li, Y.; Zhang, H.; Yang, X.; Wu, H.; Jiang, Z. Enhancement of Proton Conduction at Low Humidity by Incorporating Imidazole Microcapsules into Polymer Electrolyte Membranes. *Adv. Funct. Mater.* **2012**, *22*, 4539–4546. [CrossRef]
4. Dong, Z.; Ma, Y.; Hayat, K.; Jia, C.; Xia, S.; Zhang, X. Morphology and release profile of microcapsules encapsulating peppermint oil by complex coacervation. *J. Food Eng.* **2011**, *104*, 455–460. [CrossRef]
5. Bonilla, E.; Azuara, E.; Beristain, C.; Vernon-Carter, E.J. Predicting suitable storage conditions for spray-dried microcapsules formed with different biopolymer matrices. *Food Hydrocoll.* **2010**, *24*, 633–640. [CrossRef]
6. An, S.; Lee, M.W.; Yarin, A.L.; Yoon, S.S. A review on corrosion-protective extrinsic self-healing: Comparison of microcapsule-based systems and those based on core-shell vascular networks. *Chem. Eng. J.* **2018**, *344*, 206–220. [CrossRef]
7. Fu, J.; Song, L.; Liu, Y.; Bai, C.; Zhou, D.; Zhu, B.; Wang, T. Improving oxidative stability and release behavior of docosahexaenoic acid algae oil by microencapsulation. *J. Sci. Food Agric.* **2020**, *100*, 2774–2781. [CrossRef]
8. Yue, H.; Qiu, B.; Jia, M.; Liu, J.; Wang, J.; Huang, F.; Xu, T. Development and optimization of spray-dried functional oil microcapsules: Oxidation stability and release kinetics. *Food Sci. Nutr.* **2020**, *8*, 4730–4738. [CrossRef]
9. Martins, I.M.; Barreiro, M.F.; Coelho, M.; Rodrigues, A.E. Microencapsulation of essential oils with biodegradable polymeric carriers for cosmetic applications. *Chem. Eng. J.* **2014**, *245*, 191–200. [CrossRef]
10. Dadi, D.W.; Emire, S.A.; Hagos, A.D.; Eun, J.-B. Physical and Functional Properties, Digestibility, and Storage Stability of Spray- and Freeze-Dried Microencapsulated Bioactive Products from *Moringa stenopetala* Leaves Extract. *Ind. Crops Prod.* **2020**, *156*, 112891–112991. [CrossRef]
11. Ravanfar, R.; Comunian, T.A.; Abbaspourrad, A. Thermoresponsive, water-dispersible microcapsules with a lipid-polysaccharide shell to protect heat-sensitive colorants. *Food Hydrocoll.* **2018**, *81*, 419–428. [CrossRef]
12. Albadran, H.A.; Chatzifragkou, A.; Khutoryanskiy, V.V.; Charalampopoulos, D. Stability of probiotic Lactobacillus plantarum in dry microcapsules under accelerated storage conditions. *Food Res. Int.* **2015**, *74*, 208–216. [CrossRef] [PubMed]

13. Cortes, U.A.B.; Gutiérrez, M.C.; Mendoza, D.G.; Salitre, L.G.; Vargas, A.S.; Catzim, C.E.A.; Durán, C.C.; Valenzuela, B.E.L. Retraction notice to: "Microencapsulation and antimicrobial activity of extract acetone-methanol of *Hibiscus sabdariffa* L. using a blend modified starch and pectin as a wall material". *Ind. Crops Prod.* **2021**, *170*, 113725–113732. [CrossRef]
14. GB/T 5009.3-2016; Determination of Moisture in Foods. First Method-Direct Drying Method. National Quality and Standards Publishing & Media Co., Ltd.: Beijing, China, 2016.
15. He, L.; Hu, J.; Deng, W. Preparation and application of flavor and fragrance capsules. *Polym. Chem.* **2018**, *9*, 4926–4946. [CrossRef]
16. GB/T 5009.3-2016; Determination of Moisture in Foods. Fourth Method-Karl Fischer Method. National Quality and Standards Publishing & Media Co., Ltd.: Beijing, China, 2016.
17. Fraj, J.; Petrović, L.; Đekić, L.; Budinčić, J.M.; Bučko, S.; Katona, J. Encapsulation and release of vitamin C in double W/O/W emulsions followed by complex coacervation in gelatin-sodium caseinate system. *J. Food Eng.* **2021**, *292*, 110353–110360. [CrossRef]
18. Wang, H.; Zheng, H.; Xiao, M.; Jiang, F.T.; Fu, R.R.; Zhen, Z.H.; Xie, J.; Zhan, J.B.; Wang, X.; Zhang, L.; et al. Comparation of determination methods for moisture content in capsules. *Flavour Frag. Cosmet.* **2020**, *12*, 1–7.
19. Mehrotra, R. Infrared Spectroscopy, Gas Chromatography/Infrared in Food Analysis. In *Encyclopedia of Analytical Chemistry*; John Wiley & Sons, Ltd.: Hoboken, NJ, USA, 2006. [CrossRef]
20. Passos, C.P.; Petronilho, S.; Serôdio, A.F.; Neto, A.C.M.; Torres, D.; Rudnitskaya, A.; Nunes, C.; Kukurová, K.; Ciesarová, Z.; Rocha, S.M.; et al. HS-SPME Gas Chromatography Approach for Underivatized Acrylamide Determination in Biscuits. *Foods* **2021**, *10*, 2183. [CrossRef]
21. Lopes, G.R.; Passos, C.P.; Petronilho, S.; Rodrigues, C.; Teixeira, J.A.; Coimbra, M.A. Carbohydrates as targeting compounds to produce infusions resembling espresso coffee brews using quality by design approach. *Food Chem.* **2020**, *344*, 128613. [CrossRef]
22. Xie, W.-Q.; Chai, X.-S. Rapid determination of moisture content in paper materials by multiple headspace extraction gas chromatography. *J. Chromatogr. A* **2016**, *1443*, 62–65. [CrossRef]
23. Zhang, C.-Y.; Li, T.-F.; Chai, X.-S.; Xiao, X.-M.; Barnes, D. Determination of Porosity in Shale by Double Headspace Extraction GC Analysis. *Anal. Chem.* **2015**, *87*, 11072–11077. [CrossRef]
24. Zhang, C.-Y.; Chai, X.-S. A novel multiple headspace extraction gas chromatographic method for measuring the diffusion coefficient of methanol in water and in olive oil. *J. Chromatogr. A* **2015**, *1385*, 124–128. [CrossRef] [PubMed]
25. Xin, L.-P.; Chai, X.-S.; Hu, H.-C.; Barnes, D.G. A novel method for rapid determination of total solid content in viscous liquids by multiple headspace extraction gas chromatography. *J. Chromatogr. A* **2014**, *1358*, 299–302. [CrossRef] [PubMed]
26. Gras, K.; Luong, J.; Gras, R.; Shellie, R. Trace-level screening of dichlorophenols in processed dairy milk by headspace gas chromatography. *J. Sep. Sci.* **2016**, *39*, 3957–3963. [CrossRef] [PubMed]
27. Kolb, B.; Ettre, L.S. *Static Headspace-Gas Chromatography: Theory and Practice*, 2nd ed.; John Wiley & Sons: Hoboken, NJ, USA, 2006.
28. Hu, H.-C.; Tian, Y.-X.; Jin, H.-J.; Chai, X.-S.; Barnes, D.G. A New Headspace Gas Chromatographic Method for the Determination of Methanol Content in Paper Materials Used for Food and Drink Packaging. *J. Agric. Food Chem.* **2013**, *61*, 9362–9365. [CrossRef] [PubMed]
29. Zhang, S.-X.; Chai, X.-S.; Huang, B.-X.; Mai, X.-X. A robust method for determining water-extractable alkylphenol polyethoxylates in textile products by reaction-based headspace gas chromatography. *J. Chromatogr. A* **2015**, *1406*, 94–98. [CrossRef]
30. Xie, W.; Chai, X. Method for improving accuracy in full evaporation headspace analysis. *J. Sep. Sci.* **2014**, *40*, 1974–1978. [CrossRef]
31. Dai, Y.; Zhu, C.; Xue, M.; Chai, X.-S.; Chen, C.; Chen, R.; Hu, H. A rapid screening method for evaluating the total migratable hydrocarbons in paper products by headspace gas chromatography. *RSC Adv.* **2019**, *9*, 10226–10230. [CrossRef]
32. Chai, X.S.; Zhu, J.Y. Simultaneous measurements of solute concentration and Henry's constant using multiple headspace extraction gas chromatography. *Anal. Chem.* **1998**, *70*, 3481–3487. [CrossRef]
33. MacDougall, D.; Crummett, W.B. Guidelines for data acquisition and data quality evaluation in environmental chemistry. *Anal. Chem.* **1980**, *52*, 2242–2249. [CrossRef]
34. Gluck, S.J.; Martin, E.J. Extended Octanol-Water Partition Coefficient Determination by Dual-Mode Centrifugal Partition Chromatography. *J. Liq. Chromatogr.* **1990**, *13*, 3559–3570. [CrossRef]
35. Jin, H.-J.; Zhang, C.-Y.; Chai, X.-S. Determination of methanol partition coefficient in octanol/water system by a three-phase ratio variation headspace gas chromatographic method. *J. Chromatogr. A* **2022**, *1665*, 462825. [CrossRef] [PubMed]

Article

A Cellulose-Type Carrier for Intimate Coupling Photocatalysis and Biodegradation

Zhou Wan [1], Chunlin Jiao [1], Qilin Feng [1], Jue Wang [1], Jianhua Xiong [1,*], Guoning Chen [2,*], Shuangfei Wang [3] and Hongxiang Zhu [3]

[1] School of Resources, Environment and Materials, Guangxi University, Nanning 530004, China; wulizhouzhou@icloud.com (Z.W.); m13171793629@163.com (C.J.); rosalin28@126.com (Q.F.); wangjuecynthia@163.com (J.W.)
[2] Guangxi Bossco Environmental Protection Technology Co., Ltd., Nanning 530007, China
[3] Guangxi Key Laboratory of Clean Pulp & Papermaking and Pollution Control, Nanning 530004, China; wangsf@gxu.edu.cn (S.W.); zhx@gxu.edu.cn (H.Z.)
* Correspondence: happybear99@126.com (J.X.); chenguonin2@126.com (G.C.)

Abstract: Intimate coupling photocatalysis and biodegradation treatment technology is an emerging technology in the treatment of refractory organic matter, and the carrier plays an important role in this technology. In this paper, sugarcane cellulose was used as the basic skeleton, absorbent cotton was used as a reinforcing agent, anhydrous sodium sulfate was used as a pore-forming agent to prepare a cellulose porous support with good photocatalytic performance, and nano-TiO_2 was loaded onto it by a low-temperature bonding method. The results showed that the optimal preparation conditions of cellulose carriers were: cellulose mass fraction 1.0%; absorbent cotton 0.6 g; and Na_2SO_4 60 g. The SEM, EDS and XPS characterization further indicated that the nano-TiO_2 was uniformly loaded onto the cellulose support. The degradation experiments of Rhodamine B showed that the nano-TiO_2-loaded composite supports had good photocatalytic performance. The degradation rate of 1,2,4-trichlorobenzene was more than 92% after 6 cycles, and the experiment of adhering a large number of microorganisms on the carriers before and after the reaction showed that the cellulose-based carriers obtained the required photocatalytic performance and stability, which is a good cellulose porous carrier.

Keywords: carrier; cellulose; degradation; photocatalysis; 1,2,4-trichlorobenzene

1. Introduction

Intimate coupling photocatalysis and biodegradation (ICPB) technology [1,2] is an emerging processing technology that successfully combines photocatalytic technology and biological processing technology. In the meantime, it also integrates the advantages of both advanced oxidation technology and biodegradation technology [3], which has synergistic effects [4], and it has a good effect on the treatment of difficult-to-degrade pollutants.

The principle of pollutants degradation in ICPB is shown in Figure 1 [5]. Such active species with strong oxidizability as hydroxyl radical, superoxide radicals and holes [6–8], generated from a catalyst on the surface of carriers under light, decompose pollutants into simple and easy-biodegradable intermediate products. These products will transform into carbon dioxide and water by microbial metabolism in carriers, not excluding the direct degradation of pollutants. When considering studies related to tetracycline [9], 4-chlorophenol [10,11], methylene [12] and other pollutants in the ICPB system, a synergistic effect can be found among adsorption, photocatalysis and biodegradation [13]. The adsorption of pollutants by carriers enables the active species to oxidize and decompose pollutants in time, which reduces the damage to microorganisms from active species. The mineralization of intermediate products by microorganisms, in return, alleviates the competitive consumption of active species and improves the photocatalytic efficiency. Consequently, the porous carriers are crucial to the successful construction and operation of ICPB system.

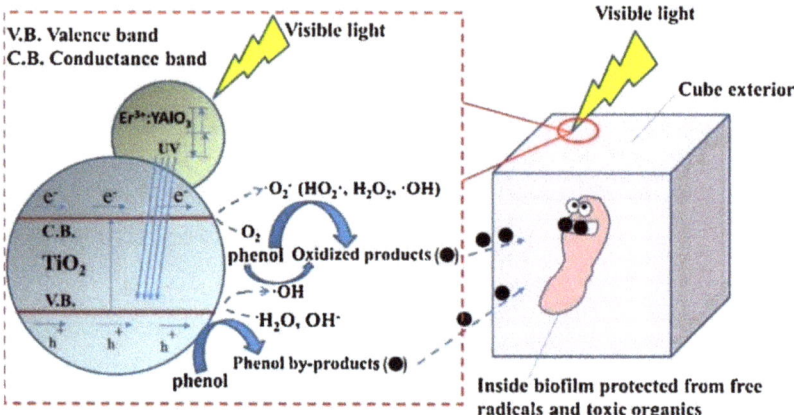

Figure 1. Schematic of the principle of pollutants degradation in ICPB.

The porous carriers involved in ICPB system mainly contain porous ceramic [14,15], cellulose [2] and the polyurethane sponge carriers [16]. Ceramic carriers have the advantage of strong stability, reusability and durability, etc. Despite this, it is difficult to ensure that carriers flow in a reactor, which makes the operational mode of the cycle between photocatalysis and microbial degradation [17]. Except for the good adsorption performance and stability [18], the lower density of polyurethane sponge-type carriers is conducive to running with the current. However, it is difficult to maintain biofilm stabilization because the hydraulic sheared and recycling process is complicated. The significant adsorption performance of cellulose carriers is favorable for attaching the catalyst and microorganism, while the cellulose material is bio-friendly and will not cause secondary pollution. However, the original structure of carriers may be damaged by microorganism degradation for a long-term operation.

Considering that the method of ICPB has great potential and research value for the degradation of persistent organic pollutants, this paper selects bagasse cellulose and absorbent cotton as major materials to prepare porous carriers; it also constructs a further ICPB system to explore the possibility of 1,2,4-trichlorobenzene (1,2,4-TCB) degradation in ICPB systems and to perfect a theoretical basis and possible practical methods for degradation of persistent organic pollutants. As a typical AOX pollutant, 1,2,4-TCB is widely present in bleaching wastewater, herbicides and other pesticide wastewater, and with stable physical and chemical properties it can exist stably in water and soil environments for a long time [19]; it is also toxic to animals, plants and humans [20]. Therefore, it is particularly important to carry out research on the degradation of 1,2,4-TCB.

2. Materials and Methods

2.1. Materials

The materials obtained were: sugarcane cellulose from Guangxi Guigang Guitang Co., Ltd.; visible light-responsive titanium dioxide (nano-TiO_2) from Liuzhou Rose Nanomaterials Technology Co., Ltd.; zinc chloride from Tianjin Ou Boke Chemical Sales Co., Ltd.; sodium sulfate from Guangdong Guanghua Sci-Tech Co, Ltd.; absorbent cotton from Nanchang Leiyi Medical Appliance Co., Ltd.; Rhodamine B (RhB) from Aladdin Reagent Co., Ltd. (Shanghai, China); and 1,2,4-TCB from Macklin. All chemicals were analytically pure. Ammonium chloride (NH_4Cl), disodium hydrogen phosphate ($Na_2HPO_4 \cdot 12H_2O$), sodium dihydrogen phosphate ($NaH_2PO_4 \cdot 2H_2O$) and magnesium sulfate ($MgSO_4 \cdot 7H_2O$)

were purchased from Guangdong Chemical Reagent Engineering Technology Research and Development Center; and calcium chloride ($CaCl_2$) and ferric chloride ($FeCl_2 \cdot 6H_2O$) were purchased from the Sinopharm Group. Activated sludge came from the research center of the Guangxi Bossco Environmental Protection Technology Co., Ltd.

2.2. Preparation of Cellulose-Type Carriers

The cellulose carriers was prepared for bagasse cellulose, absorbent cotton and sodium sulfate (Na_2SO_4) in zinc chloride ($ZnCl_2$) solution. After solidifying in deionized water and being freeze-dried, the prepared carrier had a large number of pores. The specific process was as followed: (1) mixture (100 g) stirred for 60 min at a temperature of 80 °C, which included $ZnCl_2$ (70%, wt) solution and cellulose with different mass rates in the mixture of 1%, 2%, 3%, 4%; (2) different dosages of absorbent cotton at 0.4 g, 0.5 g, 0.6 g, 0.7 g and 0.8 g were added into the mixture; (3) after stirring the mixture for 60 min at a temperature of 60 °C, 40 g, 50 g, 60 g, 70 g and 80 g of Na_2SO_4 were added, respectively; after stirring for 60 min at a temperature of 60 °C, the mixture was solidified in deionized water for 2 days and freeze-dried for 2 days at a temperature of −70 °C. The carriers with a size of 5 mm × 5 mm × 5 mms were then obtained.

Using selected water absorption, wet density, porosity and retention rates as indicators of performance, the optimal conditions for the prepared carriers were analyzed. The calculation method was as followed: absorb surface moisture by filter papers after soaking the prepared carriers in deionized water for 24 h and weighing its wet weight (m_1); measured total volume (V_1) of carriers by the drainage in cylinder (100 mL, with accuracy of 1 mL); weigh the dry weight (m_0) of the carriers after drying for 6 h at a temperature of 60 °C in a vacuum drying oven; stir the mixture of water and carriers for 60 min at a speed of 500 r/min in a beaker (1 L, with 600 mL water), in which carriers were added by the volume ratio of 1/15 (carrier/water); after stirring, measure the total volume (V_2) of the carriers again. The wet density (ρ, g/cm^3), water absorption (ω, %), porosity (ε, %) and retention rates (σ, %) were then calculated using the following Equation [21]:

$$\rho = \frac{m_1}{V_0} \quad (1)$$

$$\omega = \frac{m_1 - m_0}{m_0} \times 100\% \quad (2)$$

$$\varepsilon = \frac{m_1 - m_0}{\rho_{Aq} V_1} \times 100\% \quad (3)$$

$$\sigma = \frac{V_2}{V_1} \times 100\% \quad (4)$$

2.3. Photocatalytic Performance of Cellulose Support

A xenon lamp (XHA250W, Spectrum 200 nm–1100 nm) was used as the light source; the 15 mg/L RhB solution was placed under the lamp for 5 h, and the TiO_2-loaded carriers were added to carry out the photocatalytic degradation of RhB to detect the TiO_2 loading. Experiments of 4 cycles of degradation of RhB solution were carried out to test the reusability of the carriers.

2.4. System Construction of ICPB

The catalyst was loaded onto carriers via a simple and efficient low-temperature process on the basis of previous research [22]: dissolve 1.5 g visible light-responsive titanium dioxide (nano-TiO_2) in 15 mL solution of 0.3 g/L defused sodium and stir the mixture for 15 min; after soaking in the mixture in the first step for 10 min, bake the carriers for 120 min at a temperature of 60 °C; ultrasonically clean the nano-TiO_2-carriers in deionized water for 5 min, and repeat the process 3 times.

Activated sludge was used as the biological source and cultivated in a 2.0 L reactor. The hydraulic retention time was 24 h while the aeration rate was 0.8 L/min and the pH was 6–8. Domestication was finished after 31 days and increased by one gradient every three days with the concentration of 1,2,4-TCB from 0 mg/L to 18 mg/L. The prepared carriers were added to the activated sludge for microorganisms to attach [23]. Medium composition was as follows: NH_4Cl (35.80 mg/L); $Na_2HPO_4·12H_2O$ (10.17 mg/L); $NaH_2PO_4·2H_2O$ (5.03 mg/L); $MgSO_4·7H_2O$ (2.00 mg/L); $CaCl_2$ (2.00 mg/L); $FeCl_2·6H_2O$ (1.00 mg/L).

The schematic diagram for the system of ICPB is shown in Figure 2. Using a xenon lamp (XHA250W) as a light source, place a quartz beaker (500 mL) containing 1,2,4-TCB solution at 15 cm of the xenon lamp, 300 mL of 1,2,4-TCB solution, and an initial concentration of 8.0 mg/L. The carrier dosage (volume ratio) is 8%, the pH is 5, the stirring speed is 100 r/min, and the reaction time is 7 h.

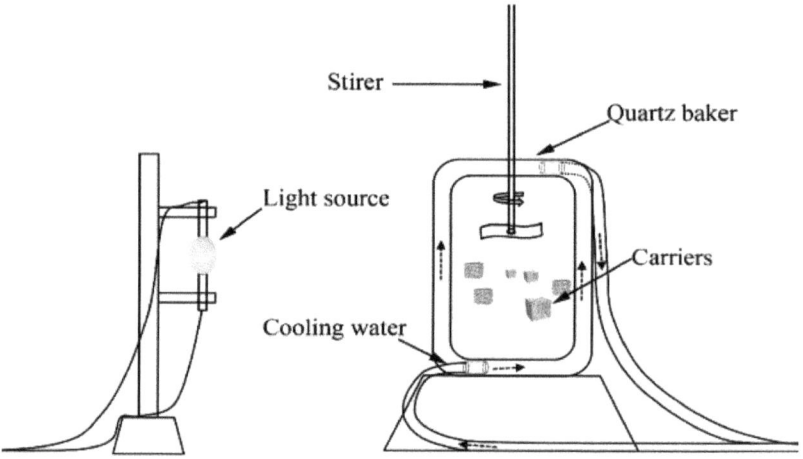

Figure 2. Schematic of the experimental setup.

2.5. Characterization

Scanning Electron Microscopy (SEM, Hitachi, Tokyo, Japan) and Energy Dispersive Spectrometry (EDS, Phenom, ThermoFisher, Waltham, MA, USA) were used to investigate morphology and surface properties. The analysis of chemical composition and electronic properties were demonstrated by X-ray Photoelectron Spectroscopy (XPS, ThermoFisher, Waltham, MA, USA), with a b-monochromatic Alkα source (hv = 1486.6 eV, 15 mA, 15 kV).

3. Results and discussion

3.1. Effect of Different Mass Fraction of Cellulose on Carrier Performance

The effects of mass fraction of cellulose on the carriers' water absorption, wet density and porosity are shown in Figure 3a,b. The values of wet density, water absorption and porosity were 0.89 g/cm^3, 513% and 87.14%, respectively, when mass fraction of cellulose was 1%. With the increasing of mass fraction of cellulose, all the values decreased gradually to 39.3%, 35.0% and 28.2%, respectively, compared with mass fraction of 4% to 1%. Cellulose is the basic framework of the carriers, and its fluffy internal structure plays an important role in forming sufficient pores in the carriers [24], which will effectively prevent the collapse of pores and being squeezed by the surrounding non-solidified solution. An appropriate amount of cellulose in carriers where mass fraction is 1% in this paper ensures more internal pores. Bagasse cellulose is hydrophilic [25], and a large amount of water can form hydrogen bonds with cellulose, effectively enhancing the ability to adsorb and store water. While the amount of cellulose increases gradually, especially mass fraction of 4%, the internal structure will become tighter and the wall of pores will become thicker.

Thus, the pores inside the carriers take less remaining space correspondingly, leading to a decrease in the water absorption and porosity of carriers.

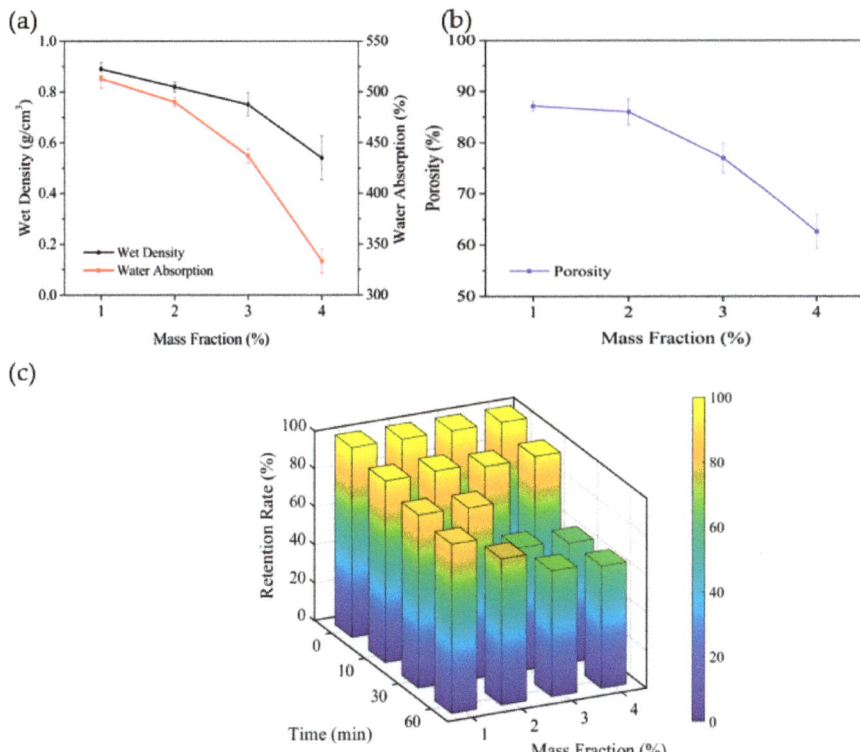

Figure 3. Effect of cellulose on water absorption: wet density (**a**), porosity (**b**), and retention (**c**) of the carrier.

The change in retention rates of carriers with different mass fraction and test times is shown in Figure 3c. Within the test time of 10 min, the retention rate of carriers reaches 100%, whatever the type of different mass fraction of cellulose. The retention rate of carriers with mass fraction of 1% is 90.2% when the test time is up to 60 min, while that of others is less than 90.0%. Dissolving the efficiency of cellulose in $ZnCl_2$ solution probably decreases gradually due to the increasing amount of cellulose. More undissolved cellulose leads to forming cellulose particles and agglomeration inside the carriers, breaking the stability of three-dimensional-net structure waved by cellulose and absorbent cotton and producing unbalanced force [26]. Therefore, the retention rate of carriers declines gradually with the increase in the mass fraction of cellulose. Instead of a high proportion, cellulose by the appropriate proportion of 1% interweaves with absorbent cotton to form a uniform three-dimensional mesh structure, with a strong ability to resist shear forces to achieve a higher retention rate of 90.2%. Therefore, the optimal mass fraction of cellulose is 1%.

3.2. Effect of Different Dosages of Absorbent Cotton on Carrier Performance

The effects of different dosages of absorbent cotton on the carriers' performances are shown in Figure 4. With the amount of absorbent cotton from 0.4 g to 0.6 g, the values of the wet density, water absorption and porosity have a bit change and maintain the variation between 0.89 g/cm^3 and 0.93 g/cm^3, 512.0% and 520.0%, 87.0% and 90.0%, respectively. The amount of absorbent cotton has a further improvement to 0.8 g, while all of the values get a significant decline by 18.0%, 11.5% and 16.2% compared to the dosage of 0.6 g. This phenomenon is explained similarly to cellulose. The increasingly absorbent cotton cannot be sufficiently dissolved, contributing to cotton aggregation and destroying the three-dimensional mesh structure of the carriers [27], which leads to a decrease in porosity and water absorption.

Figure 4c shows the change in carrier retention rate at different dosages of absorbent cotton with test time from 0–60 min. The retention rates of carriers at dosages of 0.4 g and 0.5 g gradually decrease to 64.8% and 73.4% after testing for 60 min, while the retention rates remain more than 90.0% when the dosages vary from 0.6 g to 0.8 g. The higher the dosages of absorbent cotton are, the higher the strength of cellulose carrier is [27]. Although the strength of carriers is improved with a large dosage of cotton, the number of pores would decrease because the more compact structure of carriers and three-dimensional mesh structures will be destroyed by the undissolved cotton. Therefore, to ensure adequate porosity, the optimal dosage of absorbent cotton in this paper is 0.6 g.

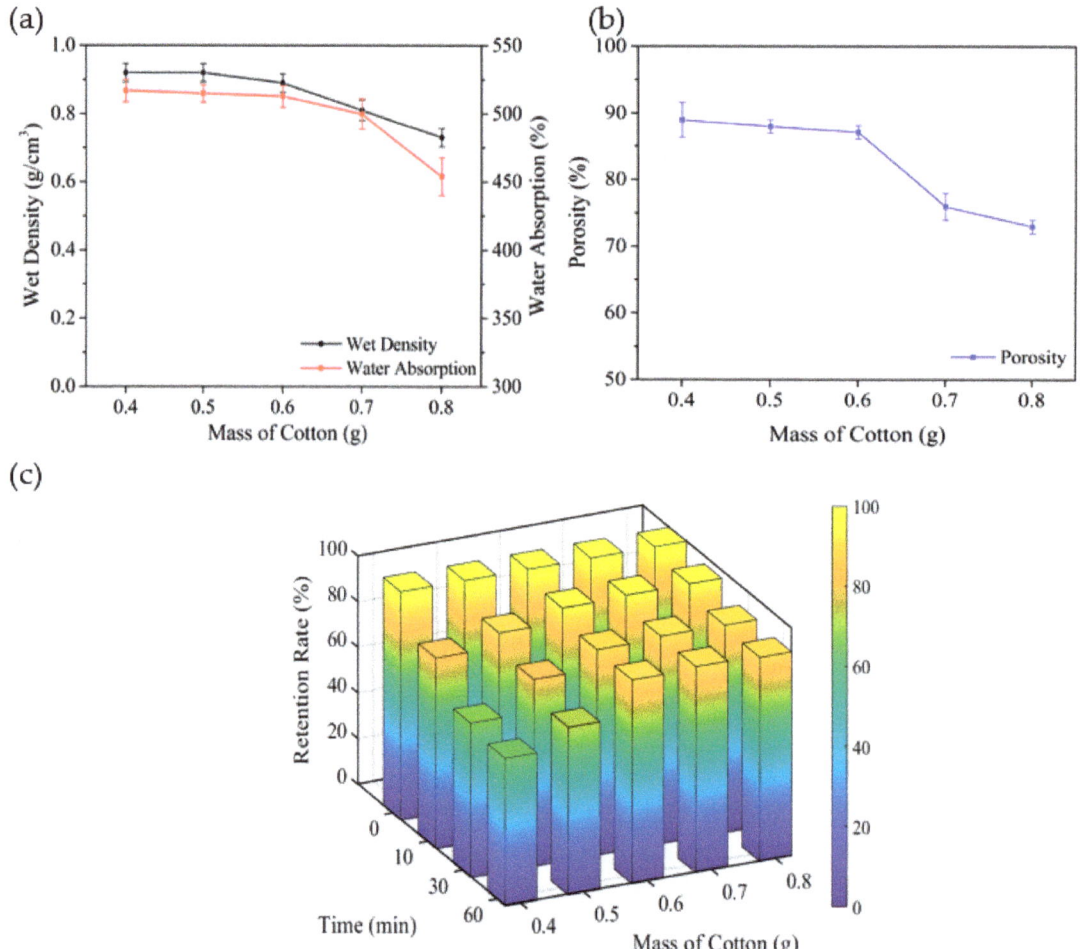

Figure 4. Effect of cotton on water absorption: wet density (**a**), porosity (**b**), and retention (**c**) of the carrier.

3.3. Effect of Different Dosages of Na_2SO_4 on Carrier Performance

As shown in Figure 5, the values of wet density, water absorption and porosity of carriers, being exactly 0.89 g/cm^3, 513% and 87.14%, respectively, reach a maximum when the dosage of Na_2SO_4 is 60 g. As a contributor of pores, Na_2SO_4 affects the number of pores in the carriers to a certain extent [28]. Theoretically, the more Na_2SO_4 is used, the fuller the porous structure and higher porosity in carriers will be, which is consistent with the evidence in Figure 5 when dosage varies from 40 g to 60 g. In addition, with the increase in Na_2SO_4 dosage, porosity and pore size, more hydrogen bonds are formed with water molecules and cellulose, to improve water absorption and wet density. This study is consistent with previous research results [26]. When the dosage is more than 60 g, the wet density and water absorption rate change little, and the porosity decreases slightly. This phenomenon comes from the porous collapse during freeze drying, when the Na_2SO_4 occupies more space in the carriers and the relatively thin supporting pore wall will collapse [27].

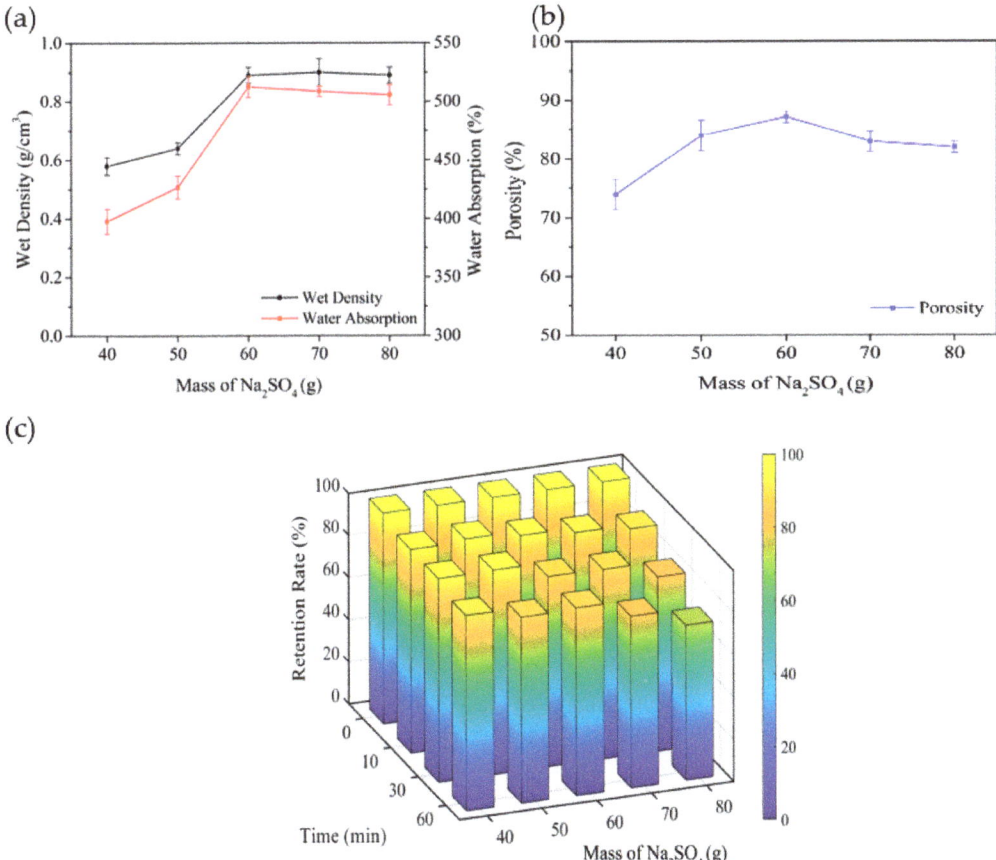

Figure 5. Effect of Na_2SO_4 on water absorption: wet density (**a**), porosity (**b**), and retention (**c**) of the carriers.

The retention rate of carriers still maintains a value more than 90.00% after testing for 60 min with the dosage of Na_2SO_4 increasing from 40 g to 60 g. With the addition of 70 g and 80 g, the retention rated decreases by 9.8% and 18.6% compared to that of 60 g. Carriers possess less pores and a thicker porous wall that makes the structure more compact, and a large number of hydrogen bonds forms into cellulose and cotton appearing with the dosage of more than 60 g, which gives the carriers a stronger ability against hydraulic shear forces. With a dosage of less than 60 g, the probability of collapse happening rises significantly on account of the porous wall becoming thinner, making the retention rate drop. Therefore, the dosage of 60 g is the optimal one for carriers.

3.4. Performance of TiO_2-Coated Cellulose-Type Carrier

The SEM images of carriers prepared for the optimal conditions are shown in Figure 6. Comparing the surface morphology of carriers before and after coating with nano-TiO_2, it can be clearly seen that a large amount of nano-TiO_2 has been coated into the carriers prepared by the method of a low-temperature process shown above, where the surface becomes rougher after coating nano-TiO_2 than in the original carriers. Additionally, whether carriers are coated with nano-TiO_2 or not, the pores of carriers are constructed with different diameters varying from 2 μm to 20 μm, indicating that the catalyst does not cover the pores of carriers and giving the possibility of growth and reproduction of microorganisms in the interior of the carriers. Moreover, EDS of nano-TiO_2-coated cellulose-type carriers show that the main elements on the surface titanium and oxygen element in Figure 7 and the amount of the titanium element is approximately twice as much as the oxygen element, which proves that nano-TiO_2 is successfully loaded onto the surface of carriers.

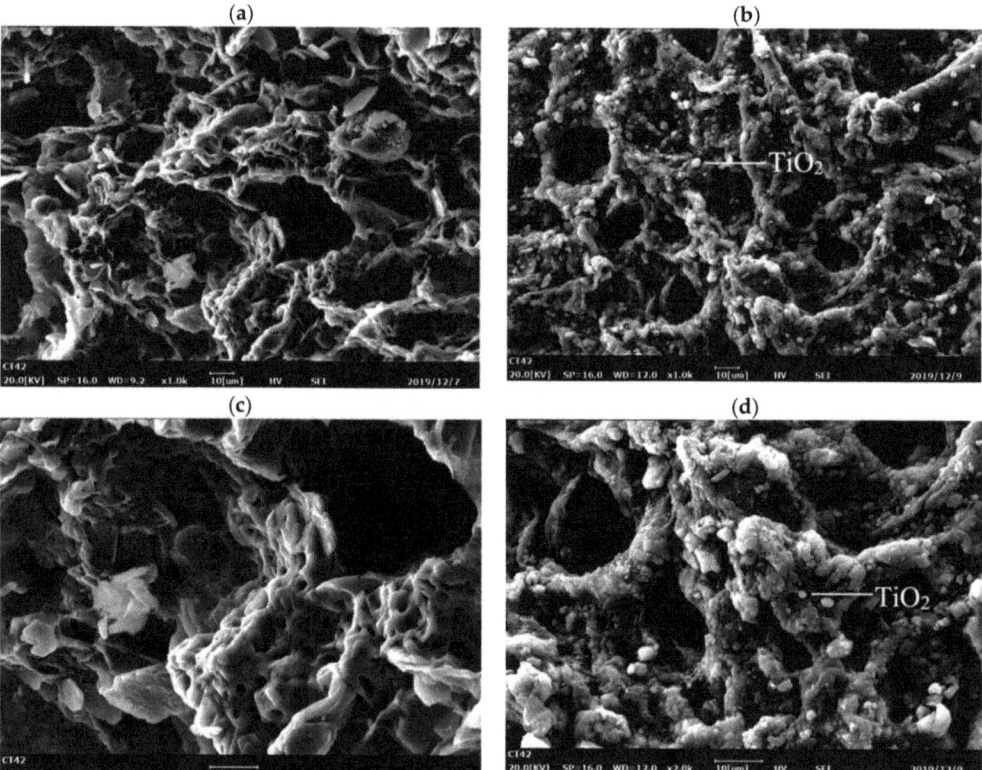

Figure 6. SEM images of carrier before and after loading nano-TiO_2: (**a,c**) with magnification of 1.0 k and 2.0 k, respectively, and without nano-TiO_2 loaded; (**b,d**) with magnification of 1.0 k and 2.0 k, respectively, and with nano-TiO_2 loaded.

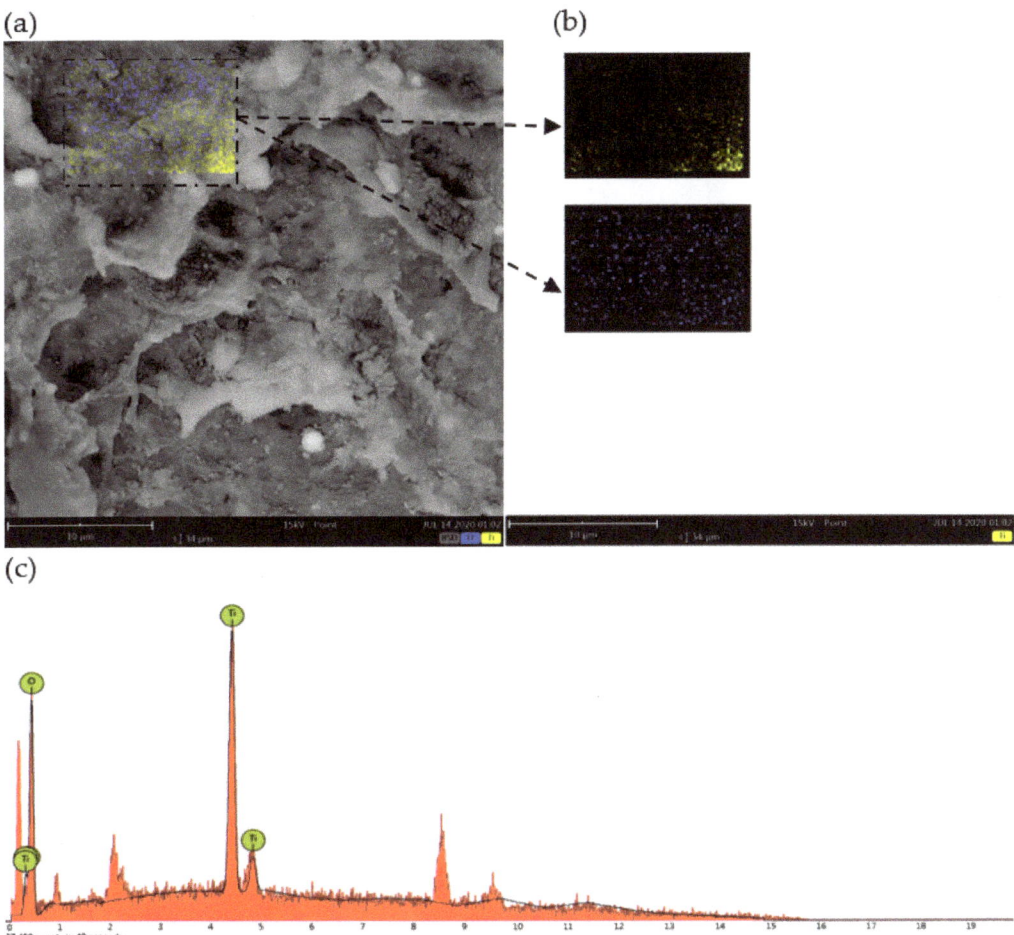

Figure 7. EDS image of nano-TiO$_2$-cellulose carrier: (**a**) Image of EDS, (**b**) Element distribution, (**c**) Energy spectrum of elements.

To further confirm the presence of phase nano-TiO$_2$ on the surface of the carriers, the chemical composition and electronic properties that were obtained from XPS analysis are shown in Figure 8. The survey spectra display the main signals from Ti, O and C (Figure 8a), and more specific properties acquired from a detailed spectrum (Figure 8b–d). Considering the composition of carriers, the C1s peak is mainly attributed to cellulose and cotton. The peaks at 532.09 eV and 532.97 eV, respectively, correspond to C–O and O–C=O bonds, associated to functional groups; for example, hydroxyl and carboxyl constructed in cellulose and cotton. The structure of the carbon skeleton is demonstrated by the bond between C–C/C–H with the energy of 284.78 eV. Except for the substrate grown catalyst, a little of the signals of adjacent to the C1s peak maybe comes from carbon contamination as the sample exposing to air [29].

Analyzing the detailed spectrum of O1s core line, it was found that the peak could be deconvoluted into three components located at 532.98 eV, 531.38 eV and 529.98 eV, respectively, which originate from the titanium oxide and oxygen-containing functional groups of carriers and the surface of catalyst. The first component, consistent with the one of C1s peak, corresponds to C–O/O–C=O bonded with functional groups of cellulose molecule. The second emergence means that a low-valence Ti oxidized has been generated in catalyst, such as Ti–OH bond (Ti hydroxide species) and Ti_xO_y. The last component originates from the bond between Ti–O combining O_2^- and Ti^{4+} in nano-TiO_2.

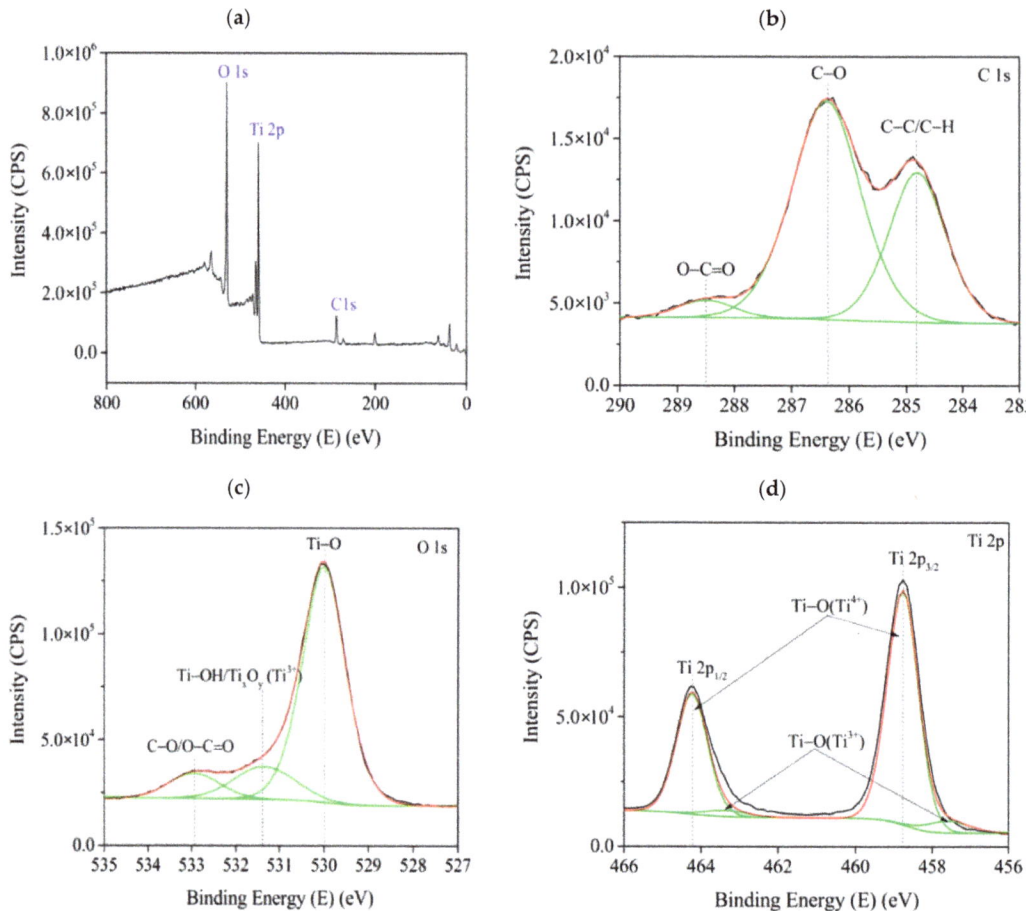

Figure 8. XPS image of TiO_2-cellulose carrier: (**a**) survey spectra, (**b**), (**c**), and (**d**) are detailed spectra of C, O, and Ti, respectively).

The detailed spectrum of Ti 2p core line has been deconvoluted into four components, including two prominent peaks of Ti 2p3/2 and Ti 2p1/2 positioned at 458.78 eV and 464.48 eV, respectively, corresponding to Ti^{4+} in titanium dioxide [30,31]. Moreover, two weaker peaks locate closely on the shoulders of prominent peaks at 457.58 eV and 463.48 eV due to the presence of oxygen vacancy and a low-valence Ti oxidized as described as lattice defects, which can improve the efficiency of photocatalysis and widen the range of excitation wavelength to the visible from the ultraviolet [32], giving a feasible explanation of visible-light reaction to the catalyst used. Therefore, the catalyst successfully loaded onto the surface of carriers.

3.5. Analysis of Photocatalytic Properties of Cellulose Composite Carriers

The above studies confirmed that nano-TiO_2 was loaded on the surface of the cellulose carriers, and the degradation experiment of methylene blue showed that titanium dioxide had good photocatalytic activity [33]. In addition, the band gap energy of TiO_2 is 3.15 eV, and this low band gap energy makes TiO_2 have a wide range of UV Vis spectra, mainly in the range of 350–600 nm; it further shows the photocatalytic activity of TiO_2. In order to further determine the photocatalytic performance of the nano-TiO_2–cellulose composite carrier and the reuse stability of the composite carrier, 15 mg/L RhB solution was used as the target pollutant for 4 cycles in this study. The experimental results are shown in Figure 9, where it can be seen from the figure that the degradation rate of RhB by the nano-TiO_2–cellulose composite carrier can reach 80.00% within 5 h. Four rounds of RhB repeated degradation experiments were carried out, and the degradation rate of RhB was basically stable at 80.00%, which proved that the nano-TiO_2–cellulose composite support had good photocatalytic performance and stability under the conditions of this study. In Table 1, compared with other materials, it is clear that the cellulose carrier loaded with TiO_2 has good photocatalytic activity, shorter time-consumption and higher degradation efficiency.

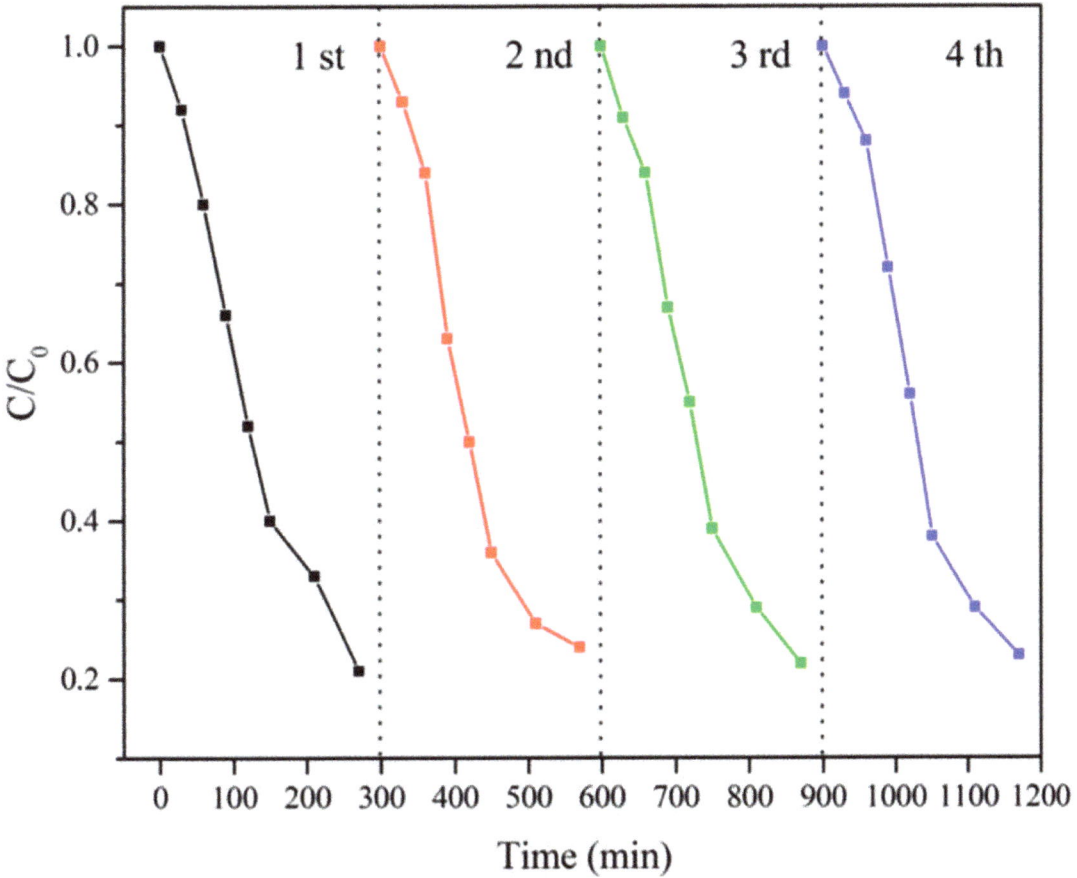

Figure 9. Cycle experiment of RhB degradation.

Table 1. The comparison of photocatalytic activity of different materials.

Carrier	Catalyst	Pollutant	Pollutant Concentration	Light Source	Time	Efficiency
/	CdS/TiO_2 [19]	1,2,4-TCB	0.1 mol/L	UV	7.5 h	32.60%
Ceramic porous carrier	TiO_2 [15]	2,4-DNT	50 mg/L	UV	60 h	78%
Sponge carrier	TiO_2 [16]	2,4,5- TCP	50 μM	UV	6 h	94.2%~98.2%
Sponge carrier	Ag/TiO_2 [34]	TCH	20 mg/L	visible	8 h	94%
Cellulose carrier	TiO_2 [12]	MB	15 mg/L	UV	6 h	92.08%

3.6. Degradation of 1,2,4-TCB in ICPB

The system of ICPB was constructed by the carrier coated catalyst and loading loaded biofilm, and 1,2,4-TCB was selected as an object for testing performance of the carriers in ICPB. The six-cycling experiments for the same batch of carriers are shown in Figure 10 with an operation time of 7 h. The degradation rate of 1,2,4-TCB in ICPB at first gets up to 95.4% and stabilizes above the level of 92.0% generally. With the carrier cycle experiment, the degradation rate gradually decreased, and the sixth decreased by 2.8% compared with the first cycle experiment; this indicates that the cellulose-type carriers can be applied to construct the system of ICPB. In addition, a slight decline in degradation rates of 1,2,4-TCB after cycles can come from falling off of a catalyst struck constantly by the stirrer, and the high degradation rate in the 2nd cycle may be related to the aggregation and accumulation of TiO_2 on the surface of a cellulose carrier. In order to confirm whether the microorganism attached to the carriers can still load on that after six-cycles running, the images of carriers were taken by SEM and are shown in Figure 11. Before the carriers participate in the degradation, a large amount of microorganism attached to the carriers (Figure 11a,b) and after the six-cycles, abundant microorganisms were still there in the carriers, which illustrates that the carriers can shelter microorganisms from the damage to active species and radiation of a light source, consistent with the research of Xiong [35]. Therefore, the cellulose-type carrier has been successfully used to construct the system of ICPB and achieve effective degradation of 1,2,4-TCB.

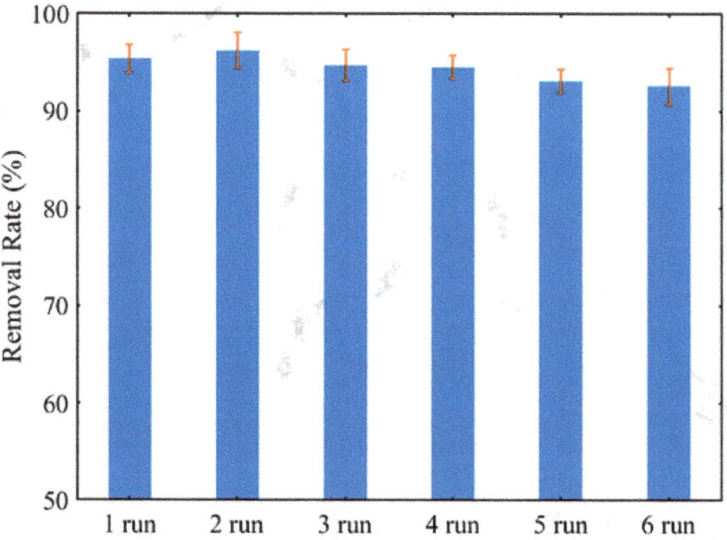

Figure 10. Change curve of 1,2,4-TrCB concentration in six consecutive batches of ICPB system.

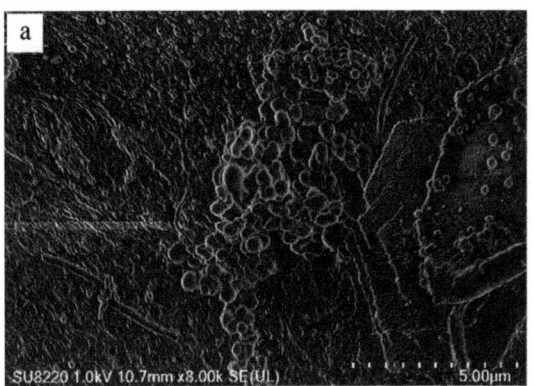 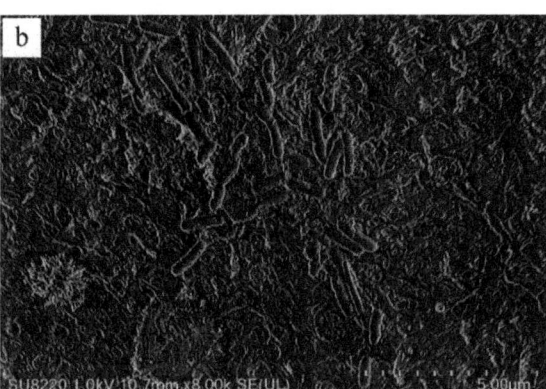

Figure 11. SEM images of biofilms located in the core of the carrier before (**a**) and after (**b**) ICPB reaction.

4. Conclusions

In this paper, sugarcane cellulose, absorbent cotton and anhydrous sodium sulfate were used as materials to prepare a cellulose porous carrier. The performance of porous the carriers were investigated by taking water absorption, wet density, porosity and retention as indicators, and the preparation process was optimized; nano-TiO$_2$ was loaded on it. The results showed that the best preparation conditions were cellulose mass fraction of 1.0%, absorbent cotton of 0.6 g, Na$_2$SO$_4$ of 60 g. The SEM, EDS and XPS characterization showed that nano-TiO$_2$ could be effectively loaded onto the surface of a cellulose carrier, and the surface and pore structure of the carriers provided conditions for microbial attachment. The degradation rate of RhB in four cycles was more than 80%, which indicates that a nano-TiO$_2$ cellulose carrier has good photocatalytic performance, which lays a good foundation for the subsequent ICPB system to achieve efficient degradation of 1,2,4-TCB.

Author Contributions: Writing—original draft preparation, Z.W.; writing—review and editing, C.J.; investigation, Q.F.; data curation, J.W.; funding acquisition, J.X.; project administration G.C.; supervision, S.W.; resources, H.Z. All authors have read and agreed to the published version of the manuscript.

Funding: This research was funded by National Natural Science Foundation of China (NSFC No: 21968005), Guangxi Major Projects of Science and Technology (Grant No. GXMPSTAB21196064), Guangxi Science and Technology Base and Special Talents (Grant No. GXSTAD19110156), National Natural Science Foundation of China (31860193), Yongjiang Project (2020013).

Institutional Review Board Statement: Not require ethical approval.

Informed Consent Statement: Not applicable.

Data Availability Statement: Not applicable.

Conflicts of Interest: The authors declare no conflict of interest.

References

1. Wang, Y.; Chen, C.; Zhou, D.; Xiong, H.; Zhou, Y.; Dong, S.; Rittmann, B.E. Eliminating partial-transformation products and mitigating residual toxicity of amoxicillin through intimately coupled photocatalysis and biodegradation. *Chemosphere* **2019**, *237*, 124491. [CrossRef] [PubMed]
2. Marsolek, M.D.; Torres, C.I.; Hausner, M.; Rittmann, B.E. Intimate coupling of photocatalysis and biodegradation in a photocatalytic circulating-bed biofilm reactor. *Biotechnol. Bioeng.* **2008**, *101*, 83–92. [CrossRef] [PubMed]
3. Ma, D.; Zou, D.; Zhou, D.; Li, T.; Dong, S.; Xu, Z.; Dong, S. Phenol removal and biofilm response in coupling of visible-light-driven photocatalysis and biodegradation: Effect of hydrothermal treatment temperature. *Int. Biodeterior. Biodegrad.* **2015**, *104*, 178–185. [CrossRef]
4. Yang, L.; Zhang, Y.; Bai, Q.; Yan, N.; Xu, H.; Rittmann, B.E. Intimately coupling of photolysis accelerates nitrobenzene biodegradation, but sequential coupling slows biodegradation. *J. Hazard. Mater.* **2015**, *287*, 252–258. [CrossRef] [PubMed]

5. Dong, S.; Dong, S.; Tian, X.; Xu, Z.; Ma, D.; Cui, B.; Rittmann, B.E. Role of self-assembly coated Er^{3+}: $YAlO_3/TiO_2$ in intimate coupling of visible-light-responsive photocatalysis and biodegradation reactions. *J. Hazard. Mater.* **2016**, *302*, 386–394. [CrossRef]
6. Kőrösi, L.; Bognár, B.; Bouderias, S.; Castelli, A.; Scarpellini, A.; Pasquale, L.; Prato, M. Highly-efficient photocatalytic generation of superoxide radicals by phase-pure rutile TiO_2 nanoparticles for azo dye removal. *Appl. Surf. Sci.* **2019**, *493*, 719–728. [CrossRef]
7. Deng, Y.; Tang, L.; Feng, C.; Zeng, G.; Wang, J.; Lu, Y.; Zhou, Y. Construction of plasmonic Ag and nitrogen-doped graphene quantum dots codecorated ultrathin graphitic carbon nitride nanosheet composites with enhanced photocatalytic activity: Full-spectrum response ability and mechanism insight. *ACS Appl. Mater. Interfaces* **2017**, *9*, 42816–42828. [CrossRef]
8. Lin, S.; Su, G.; Zheng, M.; Jia, M.; Qi, C.; Li, W. The degradation of 1,2,4-trichlorobenzene using synthesized Co_3O_4 and the hypothesized mechanism. *J. Hazard. Mater.* **2011**, *192*, 1697–1704. [CrossRef]
9. Xiong, H.; Zou, D.; Zhou, D.; Dong, S.; Wang, J.; Rittmann, B.E. Enhancing degradation and mineralization of tetracycline using intimately coupled photocatalysis and biodegradation (ICPB). *Chem. Eng. J.* **2017**, *316*, 7–14. [CrossRef]
10. Zhang, C.; Fu, L.; Xu, Z.; Xiong, H.; Zhou, D.; Huo, M. Contrasting roles of phenol and pyrocatechol on the degradation of 4-chlorophenol in a photocatalytic–biological reactor. *Environ. Sci. Pollut. Res.* **2017**, *24*, 24725–24731. [CrossRef]
11. Zhou, D.; Dong, S.; Shi, J.; Cui, X.; Ki, D.; Torres, C.I.; Rittmann, B.E. Intimate coupling of an N-doped TiO_2 photocatalyst and anode respiring bacteria for enhancing 4-chlorophenol degradation and current generation. *Chem. Eng. J.* **2017**, *317*, 882–889. [CrossRef]
12. Xiong, J.; Guo, S.; Zhao, T.; Liang, Y.; Liang, J.; Wang, S.; Chen, G. Degradation of methylene blue by intimate coupling photocatalysis and biodegradation with bagasse cellulose composite carrier. *Cellulose* **2020**, *27*, 3391–3404. [CrossRef]
13. Yu, M.; Wang, J.; Tang, L.; Feng, C.; Liu, H.; Zhang, H.; Xie, Q. Intimate coupling of photocatalysis and biodegradation for wastewater treatment: Mechanisms, recent advances and environmental applications. *Water Res.* **2020**, *175*, 115673. [CrossRef] [PubMed]
14. Xing, Z.; Zhou, W.; Du, F.; Qu, Y.; Tian, G.; Pan, K.; Fu, H. A floating macro/mesoporous crystalline anatase TiO_2 ceramic with enhanced photocatalytic performance for recalcitrant wastewater degradation. *Dalton Trans.* **2014**, *43*, 790–798. [CrossRef]
15. Wen, D.; Li, G.; Xing, R.; Park, S.; Rittmann, B.E. 2,4-DNT removal in intimately coupled photobiocatalysis: The roles of adsorption, photolysis, photocatalysis, and biotransformation. *Appl. Microbiol. Biotechnol.* **2012**, *95*, 263–272. [CrossRef]
16. Li, G.; Park, S.; Kang, D.W.; Krajmalnik-Brown, R.; Rittmann, B.E. 2,4,5-Trichlorophenol degradation using a novel TiO_2-coated biofilm carrier: Roles of adsorption, photocatalysis, and biodegradation. *Environ. Sci. Technol.* **2011**, *45*, 8359–8367. [CrossRef] [PubMed]
17. Zhang, L.; Xing, Z.; Zhang, H.; Li, Z.; Wu, X.; Zhang, X.; Zhou, W. High thermostable ordered mesoporous $SiO2-TiO_2$ coated circulating-bed biofilm reactor for unpredictable photocatalytic and biocatalytic performance. *Appl. Catal. B Environ.* **2016**, *180*, 521–529. [CrossRef]
18. Zhou, D.; Xu, Z.; Dong, S.; Huo, M.; Dong, S.; Tian, X.; Ma, D. Intimate coupling of photocatalysis and biodegradation for degrading phenol using different light types: Visible light vs UV light. *Environ. Sci. Technol.* **2015**, *49*, 7776–7783. [CrossRef]
19. Kozhevnikova, N.S.; Gorbunova, T.I.; Vorokh, A.S.; Pervova, M.G.; Zapevalov, A.Y.; Sakoutin, V.I.; Chupakhin, O.N. Nanocrystalline TiO_2 doped by small amount of pre-synthesized colloidal CdS nanoparticles for photocatalytic degradation of 1,2,4-trichlorobenzene. *Sustain. Chem. Pharm.* **2019**, *11*, 1–11. [CrossRef]
20. Carlson, A.R. Effects of lowered dissolved oxygen concentration on the toxicity of 1,2,4-trichlorobenzene to fathead minnows. *Bull Environ. Contam. Toxicol.* **1987**, *38*, 667–673. [CrossRef]
21. Xiong, J.; Liang, Y.; Cheng, H.; Guo, S.; Jiao, C.; Zhu, H.; Chen, G. Preparation and photocatalytic properties of a bagasse cellulose-supported nano-TiO_2 photocatalytic-coupled microbial carrier. *Materials* **2020**, *13*, 1645. [CrossRef] [PubMed]
22. Zhao, T.; Cheng, H.; Liang, Y.; Xiong, J.; Zhu, H.; Wang, S.; Chen, G. Preparation of TiO_2/sponge composite for photocatalytic degradation of 2,4,6-trichlorophenol. *Water Air Soil Pollut.* **2020**, *231*, 1–14. [CrossRef]
23. Zhou, G.; Li, N.; Rene, E.R.; Liu, Q.; Dai, M.; Kong, Q. Chemical composition of extracellular polymeric substances and evolution of microbial community in activated sludge exposed to ibuprofen. *J. Environ. Manag.* **2019**, *246*, 267–274. [CrossRef] [PubMed]
24. Mohamed, M.A.; Salleh WN, W.; Jaafar, J.; Ismail, A.F.; Abd Mutalib, M.; Sani NA, A.; Ong, C.S. Physicochemical characteristic of regenerated cellulose/N-doped TiO_2 nanocomposite membrane fabricated from recycled newspaper with photocatalytic activity under UV and visible light irradiation. *Chem. Eng. J.* **2016**, *284*, 202–215. [CrossRef]
25. Chin, S.F.; Jimmy, F.B.; Pang, S.C. Fabrication of Cellulose Aerogel from Sugarcane Bagasse as Drug Delivery Carriers. *J. Phys. Sci.* **2016**, *27*, 159–168. [CrossRef]
26. Jiao, W. Preparation and Study of Cellulose foam Materials. Master's Thesis, Wuhan Textile University, Wuhan, China, 2013. Available online: https://kns.cnki.net/KCMS/detail/detail.aspx?dbname=CMFD201302&filename=1013202982.nh (accessed on 17 May 2022).
27. Mamat, H.; Kibir, B. Preparation of cellulose sponge from cellulose carbamate. *Health Environ. Res. Online (HERO)* **2012**, *63*, 1637–1642. [CrossRef]
28. Xiong, J.; Yu, S.; Zhu, H.; Wang, S.; Chen, Y.; Liu, S. Dissolution and structure change of bagasse cellulose in zinc chloride solution. *BioResources* **2016**, *11*, 3813–3824. [CrossRef]
29. Güzelçimen, F.; Tanören, B.; Çetinkaya, Ç.; Kaya, M.D.; Efkere, H.İ.; Özen, Y.; Özçelik, S. The effect of thickness on surface structure of rf sputtered TiO_2 thin films by XPS, SEM/EDS, AFM and SAM. *Vacuum* **2020**, *182*, 109766. [CrossRef]

30. Desai, N.D.; Khot, K.V.; Dongale, T.; Musselman, K.P.; Bhosale, P.N. Development of dye sensitized TiO_2 thin films for efficient energy harvesting. *J. Alloys Compd.* **2019**, *790*, 1001–1013. [CrossRef]
31. Kuhn, B.L.; Paveglio, G.C.; Silvestri, S.; Muller, E.I.; Enders, M.S.; Martins, M.A.; Frizzo, C.P. TiO_2 nanoparticles coated with deep eutectic solvents: Characterization and effect on photodegradation of organic dyes. *New J. Chem.* **2019**, *43*, 1415–1423. [CrossRef]
32. Schneider, J.; Matsuoka, M.; Takeuchi, M.; Zhang, J.; Horiuchi, Y.; Anpo, M.; Bahnemann, D.W. Understanding TiO_2 photocatalysis: Mechanisms and materials. *Chem. Rev.* **2014**, *114*, 9919–9986. [CrossRef] [PubMed]
33. Huang, W.; Cheng, H.; Feng, J.; Shi, Z.; Bai, D.; Li, L. Synthesis of highly water-dispersible N-doped anatase titania based on low temperature solvent-thermal method. *Arab. J. Chem.* **2018**, *11*, 871–879. [CrossRef]
34. Ma, Y.; Xiong, H.; Zhao, Z.; Yu, Y.; Zhou, D.; Dong, S. Model-based evaluation of tetracycline hydrochloride removal and mineralization in an intimately coupled photocatalysis and biodegradation reactor. *Chem. Eng. J.* **2018**, *351*, 967–975. [CrossRef]
35. Xiong, H.; Dong, S.; Zhang, J.; Zhou, D.; Rittmann, B.E. Roles of an easily biodegradable co-substrate in enhancing tetracycline treatment in an intimately coupled photocatalytic-biological reactor. *Water Res.* **2018**, *136*, 75–83. [CrossRef] [PubMed]

Article

Preparation and Recognition Properties of Molecularly Imprinted Nanofiber Membrane of Chrysin

Yaohui Wang [1,2,3,4], Long Li [1,2,3,4], Gege Cheng [1,2,3,4], Lanfu Li [1], Xiuyu Liu [1,2,3,4,*] and Qin Huang [1,2,3,4,*]

1. School of Chemistry and Chemical Engineering, Guangxi Minzu University, Nanning 530006, China; wwangyh970218@163.com (Y.W.); lilong19980227@163.com (L.L.); ggcheng2022@163.com (G.C.); a17876072393@163.com (L.L.)
2. Key Laboratory of Chemistry and Chemical Engineering of Forest Products, State Ethnic Affairs Commission, Nanning 530006, China
3. Guangxi Key Laboratory of Chemistry and Engineering of Forest Products, Nanning 530006, China
4. Guangxi Forest Product Chemistry and Engineering Collaborative Innovation Center, Nanning 530006, China
* Correspondence: xiuyu.liu@gxun.edu.cn (X.L.); huangqin@gxun.edu.cn (Q.H.)

Abstract: The separation and extraction of chrysin from active ingredients of natural products are of great significance, but the existing separation and extraction methods have certain drawbacks. Here, chrysin molecularly imprinted nanofiber membranes (MINMs) were prepared by means of electrospinning using chrysin as a template and polyvinyl alcohol and natural renewable resource rosin ester as membrane materials, which were used for the separation of active components in the natural product. The MINM was examined using Fourier transform infrared (FT-IR) spectroscopy, scanning electron microscopy (SEM), and thermogravimetric analysis (TGA). The adsorption performance, adsorption kinetics, adsorption selectivity, and reusability of the MINM were investigated in static adsorption experiments. The analysis results show that the MINM was successfully prepared with good morphology and thermal stability. The MINM has a good adsorption capacity for chrysin, showing fast adsorption kinetics, and the maximum adsorption capacity was 127.5 mg·g^{-1}, conforming to the Langmuir isotherm model and pseudo-second-order kinetic model. In addition, the MINM exhibited good selectivity and excellent reusability. Therefore, the MINM proposed in this paper is a promising material for the adsorption and separation of chrysin.

Keywords: chrysin; electrospinning; molecular imprinting membrane; adsorption

1. Introduction

In recent years, with the development of analytical methods (high-performance liquid chromatography, ultra-high-performance liquid chromatography, electrochromatography, etc.) and extraction techniques (membrane separation, semi-bionic extraction, high-speed countercurrent chromatography, etc.), pharmacologically active natural products have gained unprecedented popularity [1,2]. Most of them have had profound effects on our lives. Chrysin is chemically known as 5,7-dihydroxy flavone. It is a natural flavonoid and is the main bioactive component isolated from traditional *Oroxylum indicum* [3–5]. Chrysin exhibits anti-oxidative [6], anti-viral, immunomodulatory, and anti-inflammatory effects [7,8]. Numerous studies have indicated that chrysin inhibits tumor cell proliferation and induces tumor cell apoptosis, restrains tumor angiogenesis, and reverses tumor cell multi-drug resistance [9–12]. It is a natural active ingredient with an anti-tumor effect. Therefore, the extraction and utilization of chrysin are of great economic importance.

According to several reviews of the literature, methods such as high-performance liquid chromatography (HPLC) [13], column chromatography [14], chromatography [15,16], adsorption [17–19], water-methanol [20,21], and ultrasonic/microwave-assisted extraction [22–24] have been developed for the analysis and separation of chrysin in *Oroxylum indicum*. However, most methods, such as column chromatography and other traditional methods, have a low separation effect on the structural analogs of chrysin. After extraction, further separation and purification are required. An efficient, low-cost material for chrysin extraction and purification is currently lacking.

In recent years, molecular imprinting technology [25–27] has been generally used for separating and purifying the effective constituents from various natural products. Compared to other methods, the molecular imprinting method has the advantages of stronger affinity and recognition ability. However, molecularly imprinted polymers are usually prepared as a whole material, resulting in most imprinted cavities lying deep within the polymer matrix. Due to these factors, it will have the disadvantages of poor imprinted loci accessibility, incomplete removal of the template, and weaker binding capacity [28–30].

Molecularly imprinted film is an effective material used to control templates located at the surface of imprinted materials; a typical example of this surface imprinting, which is carried out by immobilizing template molecules at the surface of suitable substrates, forming thin imprinted films [31,32]. Researchers have used grafting, coating, electrostatic deposition, electrostatic spinning, and other methods to prepare molecularly imprinted membranes (MIMs) to increase membrane flux [33–36]. Electrospinning nanofiber membranes have the characteristics of a large specific surface area, high porosity, and easy modification [37–39]. They have been widely used in tissue engineering, drug delivery, catalysis, wound dressings, and other fields [40–45]. Sueyoshi et al. [46] used optically active glutamic acid (Zd-Glu and Zl-Glu) as a template molecule and cellulose acetate as the base membrane to prepare molecularly imprinted membranes by means of electrospinning. The results show that the imprinted membranes prepared by electrospinning had higher permeability and flux than molecularly imprinted membranes prepared by other methods.

In this article, a molecularly imprinted material with excellent recognition and selective absorption for chrysin was prepared. We used polyvinyl alcohol (PVA) fiber as a supporting material and rosin ester as an auxiliary material to fabricate a molecularly imprinted nanofiber membrane (MINM). A detailed examination of MINM adsorption and selective recognition was conducted through the analysis of their kinetics, their isotherms, and their selective adsorption performances.

2. Experimental Section

2.1. Materials

Chrysin (480-40-0) was purchased from Shanghai Aladdin Biochemical Technology Co., Ltd. (Shanghai, China). Chloramphenicol (56-75-7) and oxytetracycline (79-57-2) were obtained from Shanghai Maclin Biochemical Technology Co., Ltd. (Shanghai, China). Polyvinyl alcohol (PVA, 9002-89-5), N,N-dimethylformamide (DMF, 68-12-2), and acetic acid were obtained from Shanghai Maclin Biochemical Technology Co., Ltd. (Shanghai, China). Methyl alcohol (67-56-1) was obtained from Chengdu Cologne Chemicals Co., Ltd. (Chengdu, China). Methacrylic acid (MAA, 79-41-4) and azobisisobutyronitrile (AIBN, 78-67-1) were obtained from Sinopharm Chemical Reagent Co., Ltd. (Shanghai, China). Ethylene glycol dimethacrylate (EGDMA, 97-90-5) was purchased from Alfa Aesar (Qingdao, China). Ethylene glycol maleic rosinate acrylate (EGMRA) was provided by Wuzhou Sun Shine Forestry & Chemicals Co., Ltd. (Guangxi, Wuzhou, China).

2.2. Preparation of Molecularly Imprinted Membranes

2.2.1. Preparation of Molecularly Imprinted Composite Membrane

The molecularly imprinted composite membrane (MICM) was prepared by means of electrospinning after mixing the molecularly imprinted polymer (MIP) and membrane material. The precipitation polymerization method was used for the preparation of the MIP microspheres. Accurately weighing chrysin (0.0675 g) with an electronic balance (Practum124-1cn, Sartorius, Göttingen, Germany), in a 250 mL three-necked flask, chrysin was dissolved in 100 mL of methanol. MAA (0.1825 g), EGDMA (1.6848 g), EGMRA (0.4200 g), and AIBN (0.0453 g) were dissolved in the solution and used an ultrasonic cleaner (KQ-800E, Kun Shan Ultrasonic Instruments Co., Ltd., Kunshan, China) to sonicate the raw materials to fully dissolve them. Hence, nitrogen was immediately added to the mixture, and it was degassed for 10 min. A condenser, thermometer, and stirring rod were then inserted into the three-necked flask, which was placed into a 70 °C constant-temperature water bath to heat and a constant-temperature reaction for 10 h under the condition of setting the stirring rate to 50 rpm. The molecularly imprinted polymer (MIP) was collected, and the template molecules and non-polymerized compounds were extracted simply from MIP microspheres by cleaning with a methanol/acetic acid mixture (9:1, v/v). Then, it was air-dried at 60 °C in an oven (FD115, Binder, Tuttlingen, Germany) for 12 h and stored in a desiccator. The non-imprinted polymer (NIP) microspheres were also prepared with the same procedure without the template molecule added to the reaction mixture.

For the encapsulation of MIP microspheres in electrospinning nanofibers, we heated 8% polyvinyl alcohol (PVA/water, w/v) in a water bath at 90 °C for 1 h to completely dissolve the PVA solution, then added 0.2% MIP to the methanol solution and ultrasonically dispersed it for 1 h to make it uniformly suspended in methanol. Then, PVA and MIP solutions of the same volume were mixed and stirred in a 60 °C water bath for 1 h to obtain a uniformly dispersed electrospinning solution. After that, the mixture was electrospun, the voltage was set to 20 kV, the speed was set to 1 mL·h^{-1}, and the iron plate collector was positioned 15 cm from the tip of the syringe. After 10 h of spinning, a molecularly imprinted composite membrane (MICM) was obtained. A non-imprinted composite membrane (NICM) was also prepared under the same conditions, except that the MIP in the electrospinning solution was replaced by NIP.

2.2.2. Preparation of Molecularly Imprinted Nanofiber Membranes

Molecularly imprinted nanofiber membranes (MINMs) were directly prepared by the electrospinning technique. PVA (1.2000 g), EGMRA (0.1200 g), and the template molecule chrysin (0.0375 g) were dissolved in a DMF/water solution (2:1, v/v). The mixture was continuously agitated in a closed vial for at least 2 h until no phase separation was observed. The solution was transferred to a 10 mL syringe installed with a metal needle that had an inner diameter of 0.8 mm. After that, the mixture was electrospun, the voltage was set to 20 kV, the speed was set to 1 mL·h^{-1}, and the iron plate collector was placed 15 cm from the tip of the syringe. After 10 h of spinning, the fibrous nanofibers were collected on the iron plate collector. After methanol-solvent extraction, nanofibers were inspected with UV–Vis spectrometers (UV-2700, Shimadzu, Kyoto, Japan) to determine that the template molecule chrysin was no longer detectable from the washing solvent. Afterward, these nanofibers were kept in a vacuum chamber for 24 h to eliminate trace solvents, then stored in a desiccator. In comparison, the non-imprinted nanofiber membranes (NINMs) were also spun in the same way without adding chrysin.

2.3. Characterization

The surface morphology of MINM and NINM after the samples were sprayed with gold was observed using a scanning electron microscope (SEM, SUPRA 55 Sapphire, Carl Zeiss Jena, Jena, Germany) under low vacuum conditions. Fourier transform infrared (FT-IR, MAGNA-IR550, Thermo Fisher Scientific, Waltham, MA, USA) spectra of membranes were measured with an infrared spectrometer with a wavenumber range of 4000–400 cm^{-1}. The thermogravimetric analysis (TGA) and differential thermal analysis (DTA) studies were performed using a thermal gravimetric analyzer (STA449F3, Netzsch-Gerätebau GmbH, Selb, Germany) [47]. The temperature was increased from 25 to 700 °C under a nitrogen atmosphere with a heating rate of 10 °C min^{-1}.

2.4. Adsorption Experiments

2.4.1. Adsorption Kinetics

To research the adsorption kinetics of membranes for chrysin, we dispersed 20 mg of the adsorbent samples into 20 mL of chrysin methanol solution (3.9333 mM), which was oscillated for 15, 30, 45, 60, 75, 90, 105, 120, 150, 180, 210, 240, and 300 min at room temperature. At each set time points, the concentration of chrysin in the solution was measured, and the adsorption mass of chrysin was calculated. The binding capacities of membranes were calculated according to the formula [48]:

$$Q_t = \frac{(C_0 - C_t) \times V}{m} \quad (1)$$

where C_0 (mM) is the initial chrysin concentration and C_t (mM) is the concentration of chrysin solution at time t (min). V (mL) is the volume of chrysin solution, and m (g) is the mass of membranes.

2.4.2. Adsorption Isotherm

To understand the controlling mechanisms and to quantify the maximum adsorption capacity of adsorbents, 20 mg of each of the membranes was added to 20 mL of chrysin solution with different concentrations and oscillated for 5 h, in which the initial concentrations were various (0.7867, 1.5733, 2.3600, 3.1467, and 3.9333 mM). We measured the concentration of chrysin in the solution after the adsorption was over. The equilibrium adsorption capacity was calculated using the following equation [49]:

$$Q_e = \frac{(C_0 - C_e) \times V}{m} \quad (2)$$

where C_0 (mM) is the initial chrysin concentration, and C_e (mM) is the equilibrium chrysin concentration. V (mL) is the volume of chrysin solution, and m (g) is the mass of membranes.

2.5. Adsorption Selectivity

To investigate the selectivity of membranes to chrysin, chloramphenicol and oxytetracycline were chosen as the compared molecules. The methanol solutions (3.9333 mM) of chrysin, chloramphenicol, and oxytetracycline were prepared, respectively, and then 20 mL of the solution was added to an Erlenmeyer flask, and then 20 mg of the adsorbent sample was added for 5 h at room temperature with shaking. Measure the concentration of chrysin, chloramphenicol, or oxytetracycline in the different solutions after the adsorption was over. The calculation formula of equilibrium adsorption capacity was the same as Formula (2).

2.6. Adsorption Reusability

After the adsorbent sample completed the adsorption process, the saturated sample was obtained by filtration. The filtered samples were washed with methanol and dried to obtain the regenerated MINM. The regenerated MINM was reused for the next adsorption test. Under the same conditions, the adsorption-desorption cycle was repeated 6 times, and the adsorption amount was measured and calculated each time.

2.7. Mechanical Properties

The mechanical properties of the MINM were measured by using an electromechanical universal testing machine (JDL-10000N, Yangzhou Tianfa Testing Machinery Co., Ltd., Yangzhou, China) at room temperature. The MINM was cut into strips (20 mm × 10 mm), and its thickness was measured at different locations with a micrometer. The MINM was clamped at both ends and stretched along its length at a certain tensile rate until broken. By averaging the results of three parallel experiments, the tensile strength and elongation at rupture of the MINM were determined. The tensile strength was the load per unit area when the sample was broken on the tensile machine, expressed in P, which was defined as follows [50–53]:

$$P = \frac{F}{S} \quad (3)$$

where P (MPa) is the tensile strength of the MINM, F (N) is the force on the fractured section when the MINM broke, and S (mm^2) is the area of the fracture surface of the MINM.

3. Results and Discussion

3.1. Optimization of Preparation Conditions of MINM

To investigate the effects of template molecule content, rosin ester content, and electrospinning voltage on the adsorption capacity of nanofiber membranes, different single-factor optimization experiments were carried out.

First, different MINMs were prepared by changing the content of template molecules (chrysin) in the spinning solution to determine the optimal content of template molecules, while the other preparation processes remained the same. As can be seen from Figure 1a, it could be found that with the increasing concentration of chrysin in the spinning solution, the MINM has an increased adsorption capacity to chrysin. When the usage content of chrysin was 0.25%, the optimum adsorption capacity of the MINM was successfully prepared, which should be attributed to the production of numerous imprinting cavities and recognition loci. There was, however, an obvious decrease in the adsorption ability of the MINM after increasing the chrysin content in the spinning solution continuously, which could be because the competitive synthesis loci of chrysin occurred when the amount of the chrysin was in excess, resulting in the decrease in the imprinting effect.

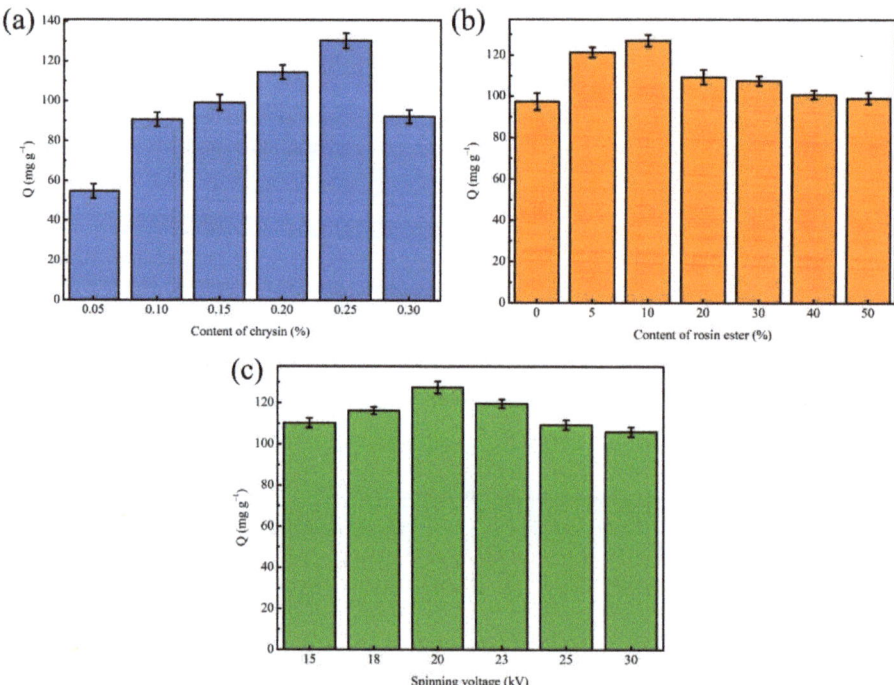

Figure 1. (a) Effect of different content of chrysin on the adsorption capacity of MINM; (b) effect of different rosin ester content on the adsorption capacity of MINM; (c) effect of different spinning voltage on the adsorption capacity of MINM.

Second, to acquire the optimal concentration of rosin ester in the preparation of the MINM, MINMs with different rosin ester contents were prepared with the other conditions unchanged. As observed in Figure 1b, the MINM prepared with different contents of rosin ester showed different adsorption capacities. The adsorption capacity of the prepared MINM increased as the content of rosin ester increased. An appropriate amount of rosin ester enhanced the rigidity and mechanical properties of the MINM and maintained the spatial structure and cavity of the MINM, thereby improving the specific binding ability to the imprinted loci. When the content of rosin ester reached 10%, the adsorption capacity reached the maximum. However, superabundant rosin ester will increase the viscosity, resulting in a decrease in the pore size of the prepared MINM and membrane flux, which will make it difficult for chrysin to reach the binding cavities in the MINM, resulting in a decrease in adsorption capacity. Therefore, the MINM with a 10% addition of rosin ester had the best adsorption capacity of all the MINMs.

Third, the effect of the electrospinning voltage on the adsorption capacity during the preparation of MINMs was then studied by ranging the voltage from 15 kV to 30 kV. It can be seen in Figure 1c that MINMs prepared with different spinning voltages exhibited different adsorption capacities. As shown, when the voltage reached 20 kV, the optimum adsorption capacity of the MINM toward chrysin was achieved; however, continuing to increase the voltage reduces the adsorption capacity. This could be because as the voltage increased, the structure of the prepared nanofiber membrane became more uniform, meaning that the surface of the membrane adsorbed chrysin more easily. In contrast, when the voltage was applied over 20 kV, the adsorption capacity of the MINM toward chrysin reduced gradually. Therefore, the optimal voltage in the spinning process was 20 kV.

3.2. Morphology of MINM

The surface characteristics of the membranes were investigated by SEM, and the results are shown in Figure 2a,c,e. The SEM photographs illustrate that the MINM and NINM had an appreciable difference in morphology and fiber diameter. The MICM had microsphere particles embedded in the fibers. The diameter distributions of the MINM, NINM, and MICM are shown in Figure 2b,d,f; the average diameter of 80 of the randomly selected nanofibers in the MINM was about 489 nm. However, the average diameter of the nanofibers in the NINM was nearly 246 nm. The average diameter of the nanofibers in the MICM was 205 nm. The MINM provides an excellent surface area and porosity to enable the transport of the chrysin molecules through the membranes. This network structure of membranes supports the easy diffusion of chrysin molecules across the membrane surface, which provides a high potential for chrysin recognition applications. According to these results, the different recognition behaviors of the MINM toward chrysin are caused by the efficient footprints and not the morphological differences.

To provide evidence for the process of the imprinting of chrysin, the MINM, NINM, and unwashed MINM were compared, as shown in Figure 2g. The FT-IR of the washed MINM and NINM exhibited the semblable shapes, which indicated that these membranes had a similar backbone. The peak at 3274 cm^{-1} in the infrared spectrum belongs to the stretching vibration of the hydroxyl group (-OH) in the PVA, and the peak at 2945 cm^{-1} belongs to the stretching vibration of the methylene group (-CH2-) in the PVA. Infrared spectroscopy analysis was performed on the unwashed MINM and the washed MINM to study the interaction between PVA and chrysin. As shown in Figure 2g, after chrysin was added, a vibration stretch peak appeared at 1617 cm^{-1}, which represents the stretch peak of the benzene ring skeleton on the chrysin, but the corresponding peak did not appear in the MINM after washing, indicating that the chrysin was successfully washed. The infrared comparison between the MINM and NINM after washing showed that there was a displacement at 1734 cm^{-1}, which may be caused by the formation of hydrogen bonds between PVA and chrysin. These results indicated that the combination of chrysin and the membrane material was successful, and chrysin was removed successfully with methanol solution. The FTIR of the MICM, the peak at 3274 cm^{-1} belongs to the stretching vibration of the hydroxyl group (-OH) in the PVA, and the peak at 2945 cm^{-1} belongs to the stretching vibration of the methylene group (-CH2-) in the PVA. The peak at 1734 cm^{-1} belongs to the stretching vibration of the carbonyl group (C=O) in the molecularly imprinted polymer. Combined with SEM and FTIR of the MICM, it can be seen that the molecularly imprinted polymer was successfully incorporated into the MICM.

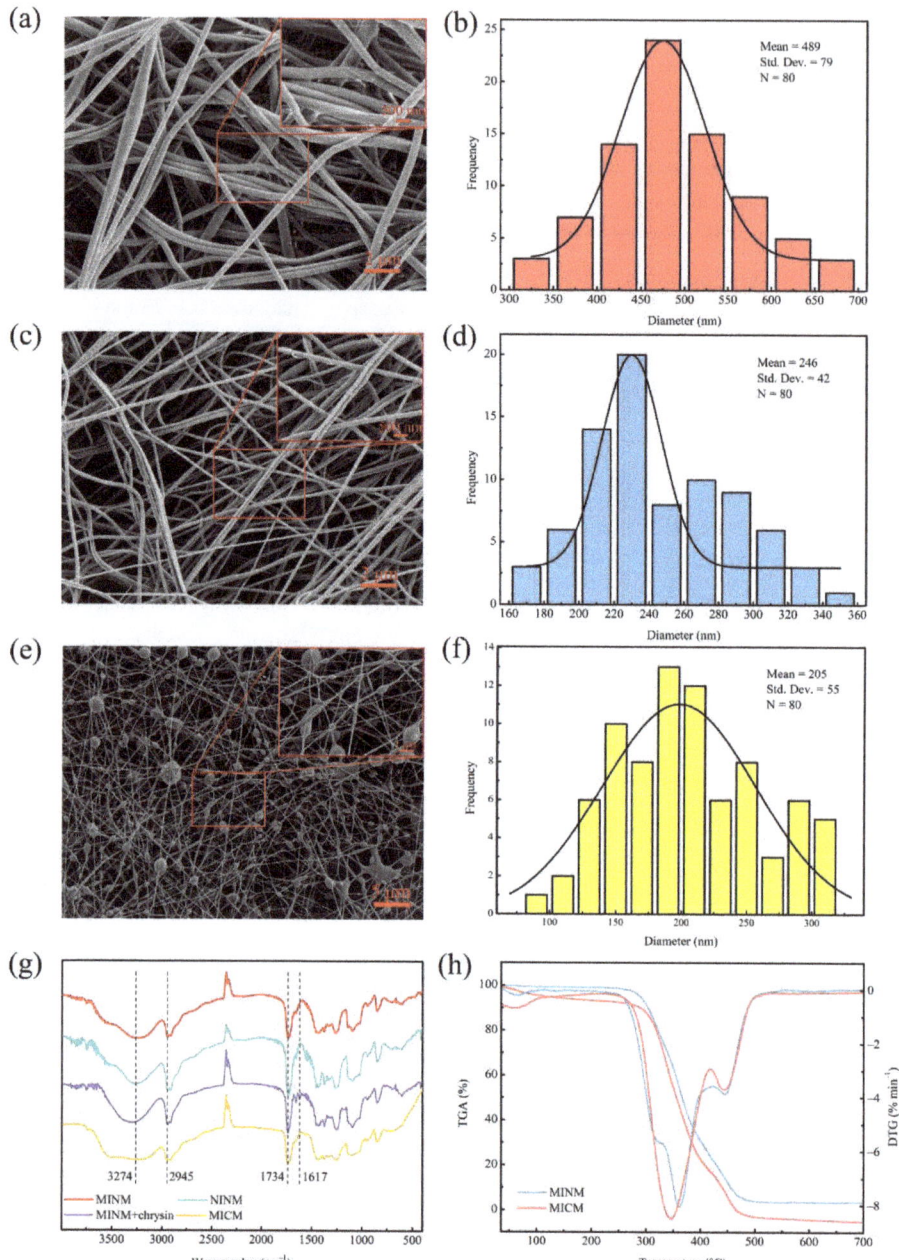

Figure 2. (**a**) SEM micrographs of MINM; (**b**) diameter distribution of MINM; (**c**) SEM micrographs of NINM; (**d**) diameter distribution of NINM; (**e**) SEM micrographs of MICM; (**f**) diameter distribution of MICM; (**g**) FT-IR spectra of the MINM, NINM, unwashed MINM, and MICM; (**h**) TGA and DTG curves of MINM and MICM.

The thermostability of the MINM and MICM was tested by TGA analysis, as shown in Figure 2h. The MINM began to decompose at approximately 275 °C, the decomposition rate reached the maximum level when the temperature was about 361 °C, and at the end of the thermal decomposition process, the temperature reached 480 °C. However, the MICM began to decompose at approximately 255 °C, the decomposition rate reached the maximum level when the temperature was about 345 °C, and at the end of the thermal decomposition process, the temperature reached 500 °C. The reason for decomposition was mainly ascribed to the scission of the main chain and the scission of cross-linked bonds. The TGA results show that the MINM had excellent thermostability. This was ascribed to the characteristic hydrocarbon-based phenanthrene rings of the EGMRA, which raised the thermostability of the MINM.

3.3. Adsorption Kinetics

In order to clarify the adsorption rate control mechanism of the adsorption process, the adsorption kinetics experiment was carried out on the MINM, NINM, MICM, and NICM under the condition of the initial concentration of chrysin solution of 3.9333 mM, and the results of the experiment are shown as Figure 3a. In the four adsorbent samples, the adsorption capacity was raised with the increase in time. The adsorption capacity of the MINM for chrysin was significantly greater than that of the NINM, MICM, and NICM, while the adsorption capacity of the MICM was significantly greater than that of the NICM. In the first 90 min, the adsorption was in the rapid adsorption stage, the adsorption capacity reached about 80% of the maximum adsorption capacity, and the adsorption reached equilibrium after 120 min, the maximum adsorption capacity of the MINM was 127.5 mg·g^{-1}, and the adsorbed chrysin amounts in this study were much higher than those reported previously [54]. It was because, in the early stage of adsorption, the concentration of chrysin was higher, and the molecular diffusion rate was faster. Numerous binding loci on the surface of the MINM and MICM quickly and specifically adsorbed chrysin. When the binding loci on the surface reached saturation, the adsorption rate gradually slowed down. There were no imprinting binding loci matching with chrysin in the NINM and NICM, so the adsorption rate was relatively slow, and the adsorption capacity to chrysin was low. In comparison to the MICM, the MINM exhibited higher adsorption capacity. because, in the MICM, the chrysin access to imprinted loci was much more difficult. Many MIP particles were not on the surface of the fibers but inside and so less accessible.

Moreover, pseudo-second-order and pseudo-first-order models were used for fitting kinetic curves to study adsorption mechanisms, such as physical adsorption and chemical adsorption [55].

The pseudo-first-order kinetic model was calculated by the equation:

$$ln(Q_e - Q_t) = lnQ_e - \frac{k_1}{2.303}t \quad (4)$$

The pseudo-second-order kinetic model was calculated by the equation:

$$\frac{t}{Q_t} = \frac{1}{k_2 Q_e^2} + \frac{t}{Q_e} \quad (5)$$

where Q_t (mg·g^{-1}) is the adsorption quantity at time t (h), Q_e is the amount of chrysin absorbed at equilibrium, and k_1 and k_2 are the equilibrium rate constant of two kinetics models.

The fitting curves of the two kinetic models are shown in Figure 3b,c. The pseudo-first-order kinetic theory predicts that adsorption loci occupancy is proportional to the unoccupied loci, whereas, in the pseudo-second-order kinetic model, the adsorption rate is determined by chemisorption between the template molecule and the adsorbent. The corresponding parameters of the two kinetic equations are determined and exhibited in Table 1.

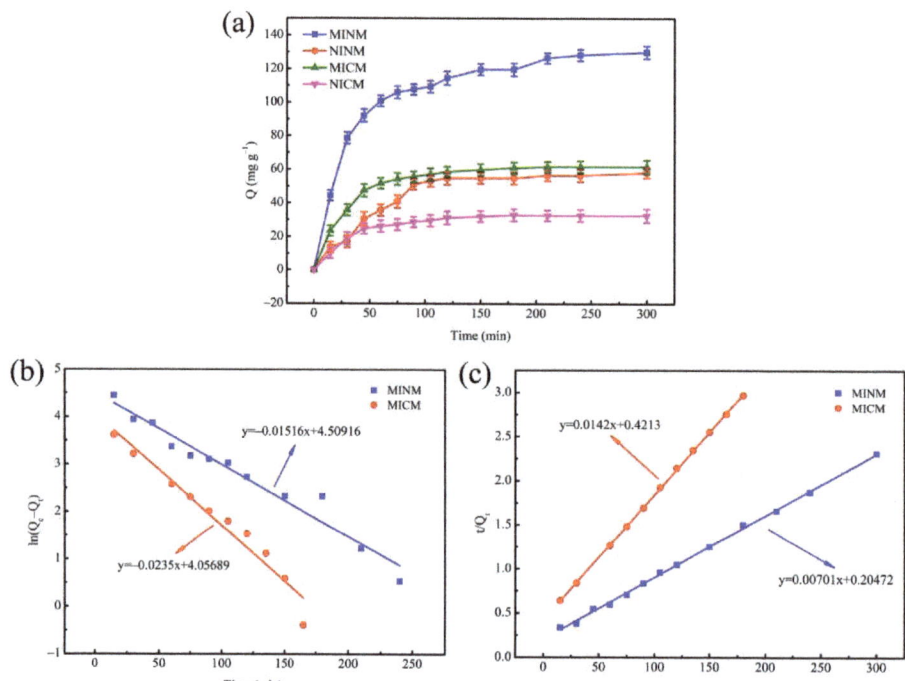

Figure 3. (a) The adsorption kinetics of MINM, NINM, MICM, and NICM; (b) the pseudo-first-order kinetic model of MINM and MICM; (c) the pseudo-second-order kinetic model of MINM and MICM.

Table 1. Kinetic data of pseudo-first-order kinetic model and pseudo-second-order kinetic model.

Samples	Pseudo-First-Order Kinetic		Pseudo-Second-Order Kinetic	
	k_1 (min^{-1})	R^2	k_2 (g·mg^{-1} min^{-1})	R^2
MINM	0.0349	0.9538	0.24×10^{-3}	0.9984
MICM	0.0541	0.9524	0.48×10^{-3}	0.9998

The correlation coefficient of the MICM's pseudo-second-order kinetic model ($R^2 = 0.9998$) is larger than that of the pseudo-first-order kinetic model ($R^2 = 0.9524$). It indicates that the adsorption process was more in line with the pseudo-second-order kinetic model, and the adsorption rate was affected by the binding ability of chrysin and the quantity of imprinting binding loci, which indicates that chemical interaction plays a leading role in the chrysin adsorption process. Similarly, the pseudo-second-order kinetic model's correlation coefficient ($R^2 = 0.9984$) of the MINM was higher than that of the pseudo-first-order kinetic model ($R^2 = 0.9538$) of the MINM. These results indicate that physical and chemical adsorption existed in the adsorption process, but chemisorption was prevailing. These results correlate with the high surface area and porosity of the MINM, the effective molecularly imprinted cavities due to chrysin on the surfaces of the MINM, and the non-covalent interaction between the MINM and chrysin.

3.4. Adsorption Isotherm

In order to measure the adsorption behavior of molecularly imprinted nanofiber membranes and molecularly imprinted composite membranes, the isotherms studies were performed, as shown in Figure 4a. With an increasing chrysin concentration, equilibrium adsorption capacity increased. It was indicated that molecularly imprinted membranes' binding capacity was better than that of non-molecularly imprinted membranes in the same situation, which implied the existence of abundant recognition loci and affinity capacity for template molecular (chrysin) on the surface of the molecularly imprinted membranes. However, under the same conditions of two kinds of molecularly imprinted membranes, the binding capacity of the MINM was significantly higher than that of the MICM. It may be because there were more recognition loci on the surface.

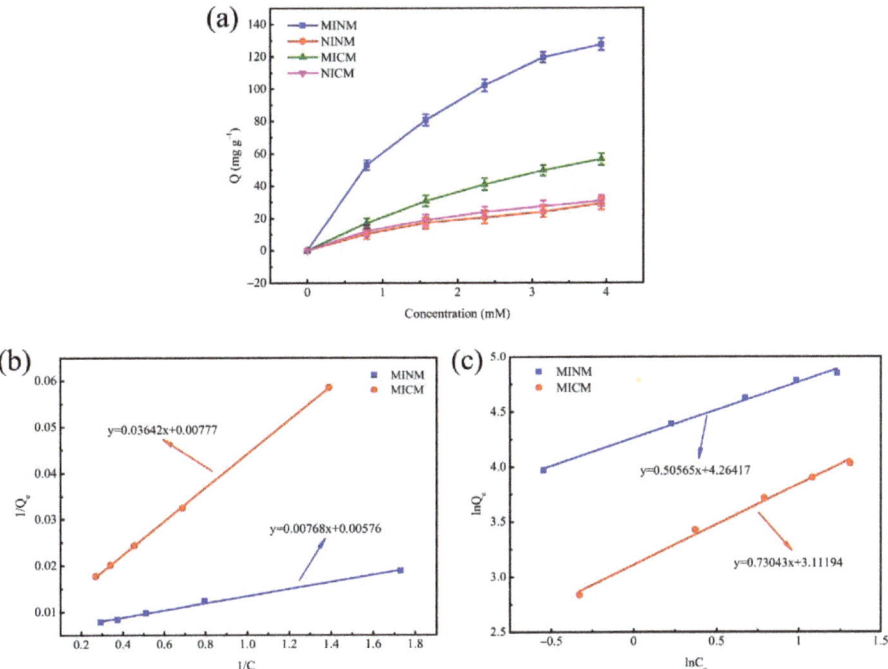

Figure 4. (a) The adsorption isotherms of MINM, NINM, MICM, and NICM; (b) the Langmuir isotherm model of MINM and MICM; (c) the Freundlich isotherm model of MINM and MICM.

In addition, adsorption equilibrium data of membranes used two typical isotherm models for adsorption [48]. The Langmuir isotherm model assumes the existence of monolayer adsorption onto a surface with a limited number of binding loci, and the Freundlich isotherm model assumes the exponential distribution of adsorption loci on the multilayer adsorption. The two isotherm models were mathematically described as follows:

Langmuir isotherm:

$$\frac{1}{Q_e} = \frac{1}{Q_m} + \frac{1}{k_3 Q_m} \times \frac{1}{C_e} \quad (6)$$

Freundlich isotherm:

$$lnQ_e = lnk_4 + \frac{1}{n} lnC_e \quad (7)$$

where Q_m (mg·g^{-1}) is the greatest adsorption quantity, Q_e (mg·g^{-1}) is the chrysin adsorption quantity at different initial concentrations, C_e (mM) is the equilibrium concentration, k is the constant, and $1/n$ is the heterogeneity factor indicating adsorption intensity.

Two kinds of adsorption isotherm for the MINM and MICM are shown in Figure 4b,c. Additionally, various kinds of isotherm parameters are shown in Table 2. It is observed that the MINM and MICM had correlation coefficients of 0.9976 and 0.9999, respectively, for the Langmuir adsorption isotherm, while MINM and MICM had correlation coefficients of 0.9921 and 0.9926, respectively, concerning the Freundlich adsorption isotherm. The results show that, in the studied concentration range, the adsorption of chrysin matches the Langmuir model better than the Freundlich model. Meanwhile, both MINM and MICM had a 1/n value in the range of 0.5–1, illustrating that the two kinds of membranes are excellent adsorption materials for chrysin.

Table 2. Parameters of Langmuir adsorption model and Freundlich adsorption model.

Samples	Langmuir Isotherm			Freundlich Isotherm		
	k_3 (mM^{-1})	R^2	Q_m (mg·g^{-1})	k_4 (mM^{-1})	R^2	1/n
MINM	0.7500	0.9976	173.611	71.1059	0.9921	0.5057
MICM	0.2133	0.9999	128.700	22.4646	0.9926	0.7304

3.5. Adsorption Selectivity

The adsorption selectivity is an essential characteristic for the application of MIMs. Thus, to examine the selectivity of MIMs to chrysin, chloramphenicol and oxytetracycline were chosen as comparative substrates in the selective adsorption test. These two molecules have similar structures and functional groups to chrysin. The selectivity research was carried out on chrysin and its comparative substrates at the concentration of 3.9333 mM. As shown in Figure 5a, the MINM had adsorption capacity for all three substances, but the adsorption capacity of the MINM for chrysin was significantly higher than for other molecules. Although the adsorption capacity of the MICM for chrysin was also better than the compared molecules, the adsorption capacity was lower than that of the MINM. The NINM and NICM had poor adsorption capacity for the three substances, and there was not much difference. All of the above results indicate that chrysin was able to selectively adsorb the MINM and MICM due to the imprinting cavities formed during the preparation in the presence of chrysin, which led to the formation of affinity binding loci along with access in the MINM and MICM. The shape, size, and functional group of these recognition loci form complementary structures to chrysin. It is profitable that the MINM and MICM have an affinity for binding chrysin. However, the NINM and NICM do not have the related loci and the recognition capability coming from the imprinting effect. It is, therefore, possible to conclude that the imprinting loci on the surface of the MINM and MICM have excellent selectivity for chrysin, and the recognition ability is provided by the imprinting loci, and compared with two printing membranes, the MINM has a stronger selective recognition ability for chrysin. The results of this study not only have better selectivity but also higher chrysin adsorption capacity compared to previous reports [56,57].

3.6. Adsorption Reusability

Besides selectivity, stability and reusability are also important indexes to evaluate the performance of the MINM. To evaluate the capacity of the MINM to be regenerated and reused, the adsorption performance after repeated cycles was investigated. After each binding experiment, the MINM was washed with methanol/acetic acid solution (9:1, v/v) to remove the adsorbed molecules. Hence, we proceeded to the next adsorption cycle. The above processes were repeated until the sorption-desorption was accomplished in six cycles. The results are illustrated in Figure 5b. The results suggest that the MINM exhibited excellent adsorption capability in all six cycles. After being recycled and reused, the MINM only lost 12.08% of adsorption capacity. This decrease may be attributed to the reduction in active binding loci following regeneration and inadequate desorption of the adsorbed chrysin molecules. It indicates that the MINM could be used repeatedly due to its stability and reusability.

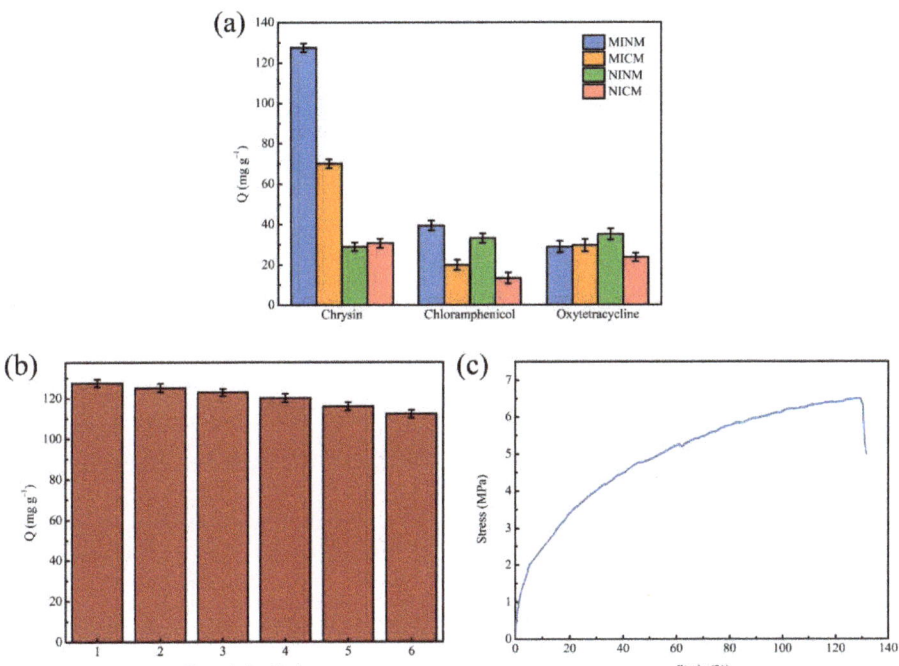

Figure 5. (**a**) The selective adsorption capacity of MINM, MICM, NINM, and NICM; (**b**) regeneration rebinding performance of MINM; (**c**) stress–strain curve of MINM.

3.7. Mechanical Properties of MINM

A special focus was placed on the mechanical properties of the membranes, since these were related to stress levels experienced during operation. The tensile strength and elongation at break of the MINM were tested, as shown in Figure 5c. The maximum tensile strength of the MINM was 6.5 MPa, and the breaking elongation of the MINM was 130%, while the maximum tensile strength of PVA nanofiber was only 1.58 MPa [58], the MINM showing excellent mechanical properties. This result indicates that this MINM enables the higher sustainability of the membrane for use in some special operating environments.

4. Conclusions

In this study, we fabricated a molecularly imprinted nanofiber membrane of chrysin using an electrospinning method for the selective adsorption of chrysin molecules. Compared with the molecularly imprinted composite membrane (MICM), the prepared MINM has a larger specific surface area, more binding sites complementary to the template are generated near the surface of the fiber, showing high adsorption capacity and significant selectivity, and the performance is generally better than that of the MICM. The adsorption kinetics suggested that the adsorption process of the MINM was more consistent with the pseudo-second-order kinetic model, illustrating that the adsorption process was controlled by chemisorption. The adsorption isotherms illustrated that the adsorption process of chrysin is in accordance with the Langmuir model rather than the Freundlich model. In addition, the MINM exhibited good thermal stability and excellent reusability. The conspicuous adsorption behavior coupled with the effortless preparation made the MINM a potential candidate for the adsorption of chrysin. In summary, we consider that this method provides a low-cost, effective way for efficiently separating and enriching the chrysin.

Author Contributions: Methodology, Y.W.; formal analysis, Y.W., X.L. and Q.H.; investigation, Y.W., X.L., Q.H. and L.L. (Long Li); data creation, Y.W., L.L. (Lanfu Li) and G.C.; writing—original draft preparation, Y.W.; writing review and editing, X.L. and Q.H.; supervision, X.L.; project administration, X.L. and Q.H. All authors have read and agreed to the published version of the manuscript.

Funding: This research received no external funding.

Institutional Review Board Statement: Not applicable.

Informed Consent Statement: Informed consent was obtained from all subjects involved in the study.

Data Availability Statement: Not applicable.

Acknowledgments: The research was supported by the National Natural Science Foundation of China (31860192), the Natural Science Foundation of Guangxi (2020GXNSFBA159009), the China Postdoctoral Science Foundation (2020M683209), the Opening Project of Guangxi Key Laboratory of Clean Pulp & Papermaking and Pollution Control (2019KF28), the Scientific Research Foundation of Guangxi Minzu University (2019KJQD10) and the Xiangsihu Young Scholars Innovative Research Team of Guangxi Minzu University (2021RSCXSHQN01).

Conflicts of Interest: The authors declare no conflict of interest.

References

1. Li, J.; Chase, H.A. Development of adsorptive (non-ionic) macroporous resins and their uses in the purification of pharmacologically-active natural products from plant sources. *Nat. Prod. Rep.* **2010**, *27*, 1493–1510. [CrossRef] [PubMed]
2. Adaskeviciute, V.; Kaskoniene, V.; Barcauskaite, K.; Kaskonas, P.; Maruska, A. The Impact of Fermentation on Bee Pollen Polyphenolic Compounds Composition. *Antioxidants* **2022**, *11*, 645. [CrossRef] [PubMed]
3. Pingili, R.B.; Pawar, A.K.; Challa, S.R.; Kodali, T.; Koppula, S.; Toleti, V. A comprehensive review on hepatoprotective and nephroprotective activities of chrysin against various drugs and toxic agents. *Chem-Biol. Interact.* **2019**, *308*, 51–60. [CrossRef] [PubMed]
4. Kasala, E.R.; Bodduluru, L.N.; Madana, R.M.; V, A.K.; Gogoi, R.; Barua, C.C. Chemopreventive and therapeutic potential of chrysin in cancer: Mechanistic perspectives. *Toxicol. Lett.* **2015**, *233*, 214–225. [CrossRef] [PubMed]
5. Kaskoniene, V.; Venskutonis, P.R. Floral Markers in Honey of Various Botanical and Geographic Origins: A Review. *Compr. Rev. Food Sci. Food Saf.* **2010**, *9*, 620–634. [CrossRef]
6. Saric, A.; Balog, T.; Sobocanec, S.; Kusic, B.; Sverko, V.; Rusak, G.; Likic, S.; Bubalo, D.; Pinto, B.; Reali, D.; et al. Antioxidant effects of flavonoid from Croatian Cystus incanus L. rich bee pollen. *Food Chem. Toxicol.* **2009**, *47*, 547–554. [CrossRef]
7. Zeinali, M.; Rezaee, S.A.; Hosseinzadeh, H. An overview on immunoregulatory and anti-inflammatory properties of chrysin and flavonoids substances. *Biomed. Pharmacother.* **2017**, *92*, 998–1009. [CrossRef]
8. Touzani, S.; Embaslat, W.; Imtara, H.; Kmail, A.; Kadan, S.; Zaid, H.; ElArabi, I.; Badiaa, L.; Saad, B. In Vitro Evaluation of the Potential Use of Propolis as a Multitarget Therapeutic Product: Physicochemical Properties, Chemical Composition, and Immunomodulatory, Antibacterial, and Anticancer Properties. *Biomed Res. Int.* **2019**, *2019*, 4836378. [CrossRef]
9. Wang, J.; Wang, H.; Sun, K.; Wang, X.; Pan, H.; Zhu, J.; Ji, X.; Li, X. Chrysin suppresses proliferation, migration, and invasion in glioblastoma cell lines via mediating the ERK/Nrf2 signaling pathway. *Drug Des. Dev. Ther.* **2018**, *12*, 721–733. [CrossRef]
10. Zhong, X.; Liu, D.; Jiang, Z.; Li, C.; Chen, L.; Xia, Y.; Liu, D.; Yao, Q.; Wang, D. Chrysin Induced Cell Apoptosis and Inhibited Invasion Through Regulation of TET1 Expression in Gastric Cancer Cells. *OncoTargets Ther.* **2020**, *13*, 3277–3287. [CrossRef]
11. Yufei, Z.; Yuqi, W.; Binyue, H.; Lingchen, T.; Xi, C.; Hoffelt, D.; Fuliang, H. Chrysin Inhibits Melanoma Tumor Metastasis via Interfering with the FOXM1/beta-Catenin Signaling. *J. Agric. Food Chem.* **2020**, *68*, 9358–9367. [CrossRef] [PubMed]
12. Chen, L.; Li, Q.; Jiang, Z.; Li, C.; Hu, H.; Wang, T.; Gao, Y.; Wang, D. Chrysin Induced Cell Apoptosis Through H19/let-7a/COPB2 Axis in Gastric Cancer Cells and Inhibited Tumor Growth. *Front. Oncol.* **2021**, *11*, 651644. [CrossRef] [PubMed]
13. Li, A.; Xuan, H.; Sun, A.; Liu, R.; Cui, J. Preparative separation of polyphenols from water-soluble fraction of Chinese propolis using macroporous absorptive resin coupled with preparative high performance liquid chromatography. *J. Chromatogr. B-Anal. Technol. Biomed. Life Sci.* **2016**, *1012–1013*, 42–49. [CrossRef]
14. Boothapandi, M.; Ravichandran, R. Antiproliferative activity of chrysin (5,7-dihydroxyflavone) from Indigofera tinctoria on human epidermoid carcinoma (A431) cells. *Eur. J. Integr. Med.* **2018**, *24*, 71–78. [CrossRef]
15. Seetharaman, P.; Gnanasekar, S.; Chandrasekaran, R.; Chandrakasan, G.; Kadarkarai, M.; Sivaperumal, S. Isolation and characterization of anticancer flavone chrysin (5,7-dihydroxy flavone)-producing endophytic fungi from *Passiflora incarnata* L. leaves. *Ann. Microbiol.* **2017**, *67*, 321–331. [CrossRef]
16. Molnar-Perl, I.; Fuzfai, Z. Chromatographic, capillary electrophoretic and capillary electrochromatographic techniques in the analysis of flavonoids. *J. Chromatogr. A* **2005**, *1073*, 201–227. [CrossRef] [PubMed]
17. Yuan, Y.; Hou, W.; Tang, M.; Luo, H.; Chen, L.-J.; Guan, Y.H.; Sutherland, I.A. Separation of Flavonoids from the Leaves of Oroxylum indicum by HSCCC. *Chromatographia* **2008**, *68*, 885–892. [CrossRef]

18. Wang, F.; Ma, X.; Qu, L. Combined Application of Macroporous Resins and Preparative High-performance Liquid Chromatography for the Separation of Steroidal Saponins from Stems and Leaves of Paris polyphylla. *Chromatographia* **2021**, *84*, 917–925. [CrossRef]
19. Cai, X.; Xiao, M.; Zou, X.; Tang, J.; Huang, B.; Xue, H. Extraction and separation of flavonoids from Malus hupehensis using high-speed countercurrent chromatography based on deep eutectic solvent. *J. Chromatogr. A* **2021**, *1641*, 461998. [CrossRef]
20. Ma, L.J.; Liu, F.; Zhong, Z.F.; Wan, J.B. Comparative study on chemical components and anti-inflammatory effects of Panax notoginseng flower extracted by water and methanol. *J. Sep. Sci.* **2017**, *40*, 4730–4739. [CrossRef]
21. Zhou, L.; Jing, T.; Zhang, P.; Zhang, L.; Cai, S.; Liu, T.; Fan, H.; Yang, G.; Lin, R.; Zhang, J. Kinetics and modeling for extraction of chrysin from Oroxylum indicum seeds. *Food Sci. Biotechnol.* **2015**, *24*, 2045–2050. [CrossRef]
22. Shen, S.; Zhou, C.; Zeng, Y.; Zhang, H.; Hossen, M.A.; Dai, J.; Li, S.; Qin, W.; Liu, Y. Structures, physicochemical and bioactive properties of polysaccharides extracted from Panax notoginseng using ultrasonic/microwave-assisted extraction. *LWT-Food Sci. Technol.* **2022**, *154*, 112446. [CrossRef]
23. Niu, Q.; Gao, Y.; Liu, P. Optimization of microwave-assisted extraction, antioxidant capacity, and characterization of total flavonoids from the leaves of Alpinia oxyphylla Miq. *Prep. Biochem. Biotechnol.* **2020**, *50*, 82–90. [CrossRef] [PubMed]
24. Wang, B.; Goldsmith, C.D.; Zhao, J.; Zhao, S.; Sheng, Z.; Yu, W. Optimization of ultrasound-assisted extraction of quercetin, luteolin, apigenin, pinocembrin and chrysin from flos populi by plackett-burman design combined with taguchi method. *Chiang Mai J. Sci.* **2018**, *45*, 427–439.
25. Yoshikawa, M.; Tharpa, K.; Dima, S.O. Molecularly Imprinted Membranes: Past, Present, and Future. *Chem. Rev.* **2016**, *116*, 11500–11528. [CrossRef]
26. Lowdon, J.W.; Dilien, H.; Singla, P.; Peeters, M.; Cleij, T.J.; van Grinsven, B.; Eersels, K. MIPs for commercial application in low-cost sensors and assays—An overview of the current status quo. *Sens. Actuators B-Chem.* **2020**, *325*, 128973. [CrossRef]
27. Ramanavicius, S.; Jagminas, A.; Ramanavicius, A. Advances in Molecularly Imprinted Polymers Based Affinity Sensors (Review). *Polymers* **2021**, *13*, 974. [CrossRef]
28. Zhao, Q.; Zhao, H.; Huang, W.; Yang, X.; Yao, L.; Liu, J.; Li, J.; Wang, J. Dual functional monomer surface molecularly imprinted microspheres for polysaccharide recognition in aqueous solution. *Anal. Methods* **2019**, *11*, 2800–2808. [CrossRef]
29. Zhang, W.; Zhang, Q.; Zhang, X.; Wu, Z.; Li, B.; Dong, X.; Wang, B. Preparation and evaluation of molecularly imprinted composite membranes for inducing crystallization of oleanolic acid in supercritical CO_2. *Anal. Methods* **2016**, *8*, 5651–5657. [CrossRef]
30. Kou, X.; Li, Q.; Lei, J.; Geng, L.; Deng, H.; Zhang, G.; Ma, G.; Su, Z.; Jiang, Q. Preparation of molecularly imprinted nanospheres by premix membrane emulsification technique. *J. Membr. Sci.* **2012**, *417–418*, 87–95. [CrossRef]
31. Liao, S.; Zhang, W.; Long, W.; Hou, D.; Yang, X.; Tan, N. Adsorption characteristics, recognition properties, and preliminary application of nordihydroguaiaretic acid molecularly imprinted polymers prepared by sol–gel surface imprinting technology. *Appl. Surf. Sci.* **2016**, *364*, 579–588. [CrossRef]
32. Xing, R.; Wang, S.; Bie, Z.; He, H.; Liu, Z. Preparation of molecularly imprinted polymers specific to glycoproteins, glycans and monosaccharides via boronate affinity controllable-oriented surface imprinting. *Nat. Protoc.* **2017**, *12*, 964–987. [CrossRef] [PubMed]
33. Cui, J.; Wu, Y.; Meng, M.; Lu, J.; Wang, C.; Zhao, J.; Yan, Y. Bio-inspired synthesis of molecularly imprinted nanocomposite membrane for selective recognition and separation of artemisinin. *J. Appl. Polym. Sci.* **2016**, *133*, 43405. [CrossRef]
34. Wu, Y.; Li, C.; Meng, M.; Lv, P.; Liu, X.; Yan, Y. Fabrication and evaluation of GO/TiO_2-based molecularly imprinted nanocomposite membranes by developing a reformative filtering strategy: Application to selective adsorption and separation membrane. *Sep. Purif. Technol.* **2019**, *212*, 245–254. [CrossRef]
35. Ying, X.; Huang, M.; Li, X. Synthesis of putrescine-imprinted double-layer nanofiber membrane by electrospinning for the selective recognition of putrescine. *J. Appl. Polym. Sci.* **2020**, *137*, 48932. [CrossRef]
36. Liu, Y.; Meng, M.; Yao, J.; Da, Z.; Feng, Y.; Yan, Y.; Li, C. Selective separation of phenol from salicylic acid effluent over molecularly imprinted polystyrene nanospheres composite alumina membranes. *Chem. Eng. J.* **2016**, *286*, 622–631. [CrossRef]
37. Xue, J.; Wu, T.; Dai, Y.; Xia, Y. Electrospinning and Electrospun Nanofibers: Methods, Materials, and Applications. *Chem. Rev.* **2019**, *119*, 5298–5415. [CrossRef]
38. Zhao, K.; Kang, S.X.; Yang, Y.Y.; Yu, D.G. Electrospun Functional Nanofiber Membrane for Antibiotic Removal in Water: Review. *Polymers* **2021**, *13*, 226. [CrossRef]
39. Agrawal, S.; Ranjan, R.; Lal, B.; Rahman, A.; Singh, S.P.; Selvaratnam, T.; Nawaz, T. Synthesis and Water Treatment Applications of Nanofibers by Electrospinning. *Processes* **2021**, *9*, 1779. [CrossRef]
40. Wang, S.; Ju, J.; Wu, S.; Lin, M.; Sui, K.; Xia, Y.; Tan, Y. Electrospinning of biocompatible alginate-based nanofiber membranes via tailoring chain flexibility. *Carbohydr. Polym.* **2020**, *230*, 115665. [CrossRef]
41. Jiang, S.; Meng, X.; Chen, B.; Wang, N.; Chen, G. Electrospinning superhydrophobic–superoleophilic $PVDF-SiO_2$ nanofibers membrane for oil–water separation. *J. Appl. Polym. Sci.* **2020**, *137*, 49546. [CrossRef]
42. Sheng, S.; Yin, X.; Chen, F.; Lv, Y.; Zhang, L.; Cao, M.; Sun, Y. Preparation and Characterization of PVA-Co-PE Drug-Loaded Nanofiber Membrane by Electrospinning Technology. *AAPS PharmSciTech* **2020**, *21*, 199. [CrossRef] [PubMed]
43. Song, X.; Li, T.; Cheng, B.; Xing, J. POSS–PU electrospinning nanofibers membrane with enhanced blood compatibility. *RSC Adv.* **2016**, *6*, 65756–65762. [CrossRef]

44. Zhao, H.; Gao, W.C.; Li, Q.; Khan, M.R.; Hu, G.H.; Liu, Y.; Wu, W.; Huang, C.X.; Li, R.K.Y. Recent advances in superhydrophobic polyurethane: Preparations and applications. *Adv Colloid. Interface.* **2022**, *303*, 102644. [CrossRef] [PubMed]
45. Gao, W.C.; Wu, W.; Chen, C.Z.; Zhao, H.; Liu, Y.; Li, Q.; Huang, C.X.; Hu, G.H.; Wang, S.F.; Shi, D.; et al. Design of a Superhydrophobic Strain Sensor with a Multilayer Structure for Human Motion Monitoring. *ACS Appl. Mater Inter.* **2022**, *14*, 1874–1884. [CrossRef]
46. Sueyoshi, Y.; Fukushima, C.; Yoshikawa, M. Molecularly imprinted nanofiber membranes from cellulose acetate aimed for chiral separation. *J. Membr. Sci.* **2010**, *357*, 90–97. [CrossRef]
47. Jiang, Y.; Wang, Z.; Zhou, L.; Jiang, S.; Liu, X.; Zhao, H.; Huang, Q.; Wang, L.; Chen, G.; Wang, S. Highly efficient and selective modification of lignin towards optically designable and multifunctional lignocellulose nanopaper for green light-management applications. *Int. J. Biol. Macromol.* **2022**, *206*, 264–276. [CrossRef]
48. Li, L.; Liu, H.; Lei, X.; Zhai, Y. Electrospun Nanofiber Membranes Containing Molecularly Imprinted Polymer (MIP) for Rhodamine B (RhB). *Adv. Chem. Eng. Sci.* **2012**, *2*, 266–274. [CrossRef]
49. Gao, J.; Zhou, S.; Hou, Z.; Zhang, Q.; Meng, M.; Li, C.; Wu, Y.; Yan, Y. One pot-economical fabrication of molecularly imprinted membrane employing carbon nanospheres sol coagulation bath with specific separation and advanced antifouling performances. *Sep. Purif. Technol.* **2019**, *218*, 59–69. [CrossRef]
50. Jiang, B.; Wang, B.; Zhang, L.; Sun, Y.; Xiao, X.; Yang, N.; Dou, H. Effect of Tween 80 on morphology and performance of poly(L-lactic acid) ultrafiltration membranes. *J. Appl. Polym. Sci.* **2017**, *134*, 44428–44436. [CrossRef]
51. Yang, F.; Ma, J.; Zhu, Q.; Qin, J. Fluorescent and mechanical properties of UiO-66/PA composite membrane. *Colloids Surf. A Physicochem. Eng. Asp.* **2021**, *627*, 127083. [CrossRef]
52. Cheng, B.-X.; Gao, W.-C.; Ren, X.-M.; Ouyang, X.-Y.; Zhao, Y.; Zhao, H.; Wu, W.; Huang, C.-X.; Liu, Y.; Liu, X.-Y.; et al. A review of microphase separation of polyurethane: Characterization and applications. *Polym. Test.* **2022**, *107*, 107489. [CrossRef]
53. Jiang, Y.; Wang, Z.; Liu, X.; Yang, Q.; Huang, Q.; Wang, L.; Dai, Y.; Qin, C.; Wang, S. Highly Transparent, UV-Shielding, and Water-Resistant Lignocellulose Nanopaper from Agro-Industrial Waste for Green Optoelectronics. *ACS Sustain. Chem. Eng.* **2020**, *8*, 17508–17519. [CrossRef]
54. Iben Nasser, I.; Algieri, C.; Garofalo, A.; Drioli, E.; Ahmed, C.; Donato, L. Hybrid imprinted membranes for selective recognition of quercetin. *Sep. Purif. Technol.* **2016**, *163*, 331–340. [CrossRef]
55. Wu, K.; Yang, W.; Jiao, Y.; Zhou, C. A surface molecularly imprinted electrospun polyethersulfone (PES) fiber mat for selective removal of bilirubin. *J. Mater. Chem. B* **2017**, *5*, 5763–5773. [CrossRef] [PubMed]
56. Liang, C.; Zhang, Z.; Zhang, H.; Ye, L.; He, J.; Ou, J.; Wu, Q. Ordered macroporous molecularly imprinted polymers prepared by a surface imprinting method and their applications to the direct extraction of flavonoids from Gingko leaves. *Food Chem.* **2020**, *309*, 125680. [CrossRef]
57. Li, X.; Dai, Y.; Row, K.H. Preparation of two-dimensional magnetic molecularly imprinted polymers based on boron nitride and a deep eutectic solvent for the selective recognition of flavonoids. *Analyst* **2019**, *144*, 1777–1788. [CrossRef]
58. Ullah, S.; Hashmi, M.; Hussain, N.; Ullah, A.; Sarwar, M.N.; Saito, Y.; Kim, S.H.; Kim, I.S. Stabilized nanofibers of polyvinyl alcohol (PVA) crosslinked by unique method for efficient removal of heavy metal ions. *J. Water Process Eng.* **2020**, *33*, 101111. [CrossRef]

Article

Removing Calcium Ions from Remelt Syrup with Rosin-Based Macroporous Cationic Resin

Gege Cheng [1,2,3,4], Wenwen Li [1,2,3,4], Long Li [1,2,3,4], Fuhou Lei [1,2,3,4], Xiuyu Liu [1,2,3,4] and Qin Huang [1,2,3,4],*

[1] School of Chemistry and Chemical Engineering, Guangxi Minzu University, Nanning 530006, China; ggcheng2022@163.com (G.C.); l19981207654321@163.com (W.L.); lilong19980227@163.com (L.L.); leifuhougxun@126.com (F.L.); xiuyu.liu@gxun.edu.cn (X.L.)
[2] Key Laboratory of Chemistry and Engineering of Forest Products, State Ethnic Affairs Commission, Nanning 530006, China
[3] Guangxi Key Laboratory of Chemistry and Engineering of Forest Products, Nanning 530006, China
[4] Guangxi Collaborative Innovation Center for Chemistry and Engineering of Forest Products, Guangxi Minzu University, Nanning 530006, China
* Correspondence: huangqin@gxun.edu.cn

Abstract: Mineral ions (mainly calcium ions) from sugarcane juice can be trapped inside the heating tubes of evaporators and vacuum boiling pans, and calcium ions are precipitated. Consequently, sugar productivity and yield are negatively affected. Calcium ions can be removed from sugarcane juice using adsorption. This paper described the experimental condition for the batch adsorption performance of rosin-based macroporous cationic resins (RMCRs) for calcium ions. The kinetics of adsorption was defined by the pseudo-first-order model, and the isotherms of calcium ions followed the Freundlich isotherm model. The maximal monolayer adsorption capacity of calcium ions was 37.05 mg·g^{-1} at a resin dosage of 4 g·L^{-1}, pH of 7.0, temperature of 75 °C, and contact time of 10 h. It appeared that the adsorption was spontaneous and endothermic based on the thermodynamic parameters. The removal rate of calcium ions in remelt syrup by RMCRs was 90.71%. Calcium ions were effectively removed from loaded RMCRs by 0.1 mol·L^{-1} of HCl, and the RMCRs could be recycled. The dynamic saturated adsorption capacity of RMCRs for calcium ions in remelt syrup was 37.90 mg·g^{-1}. These results suggest that RMCRs are inexpensive and efficient adsorbents and have potential applications for removing calcium ions in remelt syrup.

Keywords: rosin-based macroporous cationic resin; calcium ions; remelt syrup

1. Introduction

Desalination of carbonated and filtered remelt syrup is an indispensable step in the sugar industry. The inorganic minerals include calcium ions (mainly), magnesium ions, iron ions, silicic acid, phosphate, and carbonate ions in sugar juice [1]. More than 90% of the inorganic matter in the syrup relates to calcium [2]. As the concentration of sugar juice increases in the evaporators, calcium ions are precipitated as their solubilities are exceeded [3], which directly affects the productivity and achievable yield of sugar [4]. Helmut et al. [1] claimed that calcium ions in a beet juice evaporator could be reduced by 80–90% using KEBO DS (a scale inhibitor). Typically, anionic polymers, such as poly-acrylics and poly (amino polyether tetra-methylene phosphonic acid), are used in the sugar industry to inhibit calcium salts [1]. The use of scale inhibitors in the sugar industry is also limited by health concerns. These inhibitors need to be approved by relevant agencies (for example, the United States Food and Drug Administration) before they can be used in the sugar process [1]. The desalination process has been explored by many eco-friendly and green technologies, including membrane separation [5], ozonation [6], coagulation [3], and adsorption [7]. Researchers prefer adsorption because it is inexpensive, widely available, and easy to use; additionally, it has the potential to handle large-scale production [8]. The preparation and functionalization of new decalcification adsorbents for remelt syrup with high adsorption performance, environmental friendliness, and low cost has become the research focuses.

The macroporous cationic resins (MCRs) used in the syrup are mainly synthesized based on styrene and divinylbenzene [9]. Styrene and divinylbenzene have been classified as class 2B carcinogens, which are limited by health concerns, by the International Agency for Research on Cancer of the World Health Organization (IARCWHO) [10]. With the improvement of people's living standards, requirements for food quality and safety also increase. Thus, it requires us to investigate green biomass-based decalcification adsorbents that are highly adsorption-efficient, low-cost, and recyclable.

About 90% of crude rosin is rosin acid, which is derived from the exudation of conifer trees [11]. Resin acid (rosin) can be modified by an esterification and (or) addition reaction based on the conjugated double bonds and carboxyl group [12]. In the prior study, we successfully prepared a novel ethylenediamine rosin-based resin (EDAR) for the removal of phenolic compounds from water. The interaction model and adsorption mechanism of EDAR-adsorbed phenolic compounds in water were studied, which provided the basis for its application [13]. Similarly, Li et al. prepared a new rosin-based sugarcane juice decolorization agent, where the crosslinking agent was modified rosin, silica was the carrier, and the quaternary ammonium cation was the functional group [8]. The excellent mechanical properties and thermal stability of the adsorbent are due to the specific three-membered phenanthrene ring structure of rosin [14]. Thus, modified rosin has the potential to produce green, economical, and eco-friendly adsorbents in the sugar industry. However, as far as we know, rosin-based resins have not been prepared and used for removing calcium ions from remelt syrup.

In this study, rosin-based macroporous cationic resins (RMCRs) were prepared using modified rosin ethylene glycol maleic rosinate acrylate (EGMRA) as a cross-linking skeleton. The potential use of the RMCRs in removing calcium ions from remelt syrup was described. Furthermore, the mechanisms of the RMCRs adsorption of the calcium ions were elucidated by using an adsorptive isotherm and kinetic models and calculating thermodynamic parameters.

2. Materials and Methods

2.1. Materials

Hydrochloric acid (HCl, 37–39%), calcium chloride (AR), and sodium hydroxide (AR) were purchased from Sinopharm Chemical Reagent Co., Ltd. (Shanghai, China). EGMRA was provided by the Guangxi Key Laboratory of Forest Products Chemistry and Engineering (Nanning, China) [15]. The remelt syrup after carbonation and decoloration by anion resin was kindly provided by Fangcheng Sugar Refinery (Fangchenggang, China). Commercial resins, including four derivatives of the styrene-divinylbenzene copolymer (types: FPA51, FPC22 Na, FPA40 Cl, FPA90 Cl, FPC14 Na, FPC23 H, and FPC22 H), and one polymethacrylic acid (type: FPA98 Cl) were purchased from Dow Chemical (Shanghai, China).

2.2. Preparation and Characterization of RMCRS

The RMCRs used in this study were all self-prepared, and followed a previous preparation process with some modifications [16]. The functional monomers MAA (6.32 g), the porogen polypropylene glycol (1.39 g), the cross-linker EGMEA (20.25 g), and AIBN (0.2 g) were dissolved in ethyl acetate (60 mL) by sonication to obtain an organic phase. SDS (0.02 g) and PVA (0.02 g) were dissolved in deionized water in a 250 mL three-necked flask, and then the organic phase was added at 60 °C. The mixture was thermally polymerized at 80 °C for 8 h and stirred at 200 rpm. The resins were extracted with ethanol and deionized water, then immersed in 3.0% NaOH solution to ionize the COOH groups to COO− groups, and ultimately washed continuously with deionized water until the pH was approximately 7.0.

2.3. Static Adsorption and Regeneration of the Resins

The optimum adsorbent dosage, temperature, pH value, and time for calcium ions adsorption by RMCRs were determined by preliminary experiments. The adsorption of calcium ions onto the RMCRs was carried out in a conical bottle containing 50 mL of calcium ions solution and 0.200 g of RMCRs. The adsorption isotherms of calcium ions with different initial concentrations (30, 60, 90, 120, and 150 mg·L^{-1}) were obtained at temperatures of 328, 338, and 348 K. Kinetic experiments were conducted in glass flasks that contained 500 mL of calcium ions solutions at a pH of 7.0 with a calcium ions' initial concentration of 150 mg·L^{-1} and 2 g of RMCRs. The conical bottles were oscillated at 100 rpm at 348 K and sampled at regular time intervals. The concentration was determined by ICP-OES (iCAP 600 Seris, Thermo Fisher Scientific, Waltham, MA, USA). The pH value was adjusted by HCl and NaOH solutions with concentrations of 0.1 mol·L^{-1}, and the effect of pH value on calcium ion adsorption was investigated. The resins after adsorption were shaken with 0.1 mol·L^{-1} of HCl for 12 h at 298 K for regeneration. RMCRs were tested for reusability through adsorption regeneration cycles.

2.4. Fixed-Bed Column Experiments

The fixed-bed on calcium ions adsorption was conducted in silica sand glass columns (Ø1.5 × 20 cm^2) filled to depths of 4 cm with RMCRs. Under the drive of the pressure provided by a peristaltic pump, the syrup (45 °Bx, pH = 7.0) was passed through the column at a rate of 2.0 mL·min^{-1}, and the effusive calcium ions solutions were determined at various intervals.

2.5. Analysis

The RMCRs before and after the adsorption of calcium ions were characterized by Fourier-transform infrared spectroscopy (FTIR) (Nicolet 5700, Thermo Fisher Scientific, Waltham, MA, USA), X-ray diffraction (XRD) (Siemens D5000 diffractometer, Bruker, Germany), X-ray photoelectron spectroscopy (XPS) (ESCALAB 250Xi, Thermo Fisher Scientific, Waltham, MA, USA), and field emission scanning electron microscopy (FE-SEM) (JSM-7500F, JEOL, Tokyo, Japan) with energy-dispersive spectrometry (EDS). The zeta potentials of RMCRs were measured with a Zeta Potential Analyzer (Zetasizer 2000 Analyzer, Malvern, UK) at an initial pH ranging from 2.0 to 12.0.

After each experiment, the solution was filtered through 0.45 μm filters and the concentration of calcium ions was analyzed by inductively coupled plasma–atomic emission spectroscopy. All the adsorption/regeneration experiments were performed at 100 rpm with triplicates and the results were averaged from all replicates. The adsorption efficiency (q_t) and removal rate (R) were calculated as follows:

$$q_t = \frac{(C_0 - C_t) \times V}{m} \quad (1)$$

$$R(\%) = \frac{C_0 - C_t}{C_0} \times 100 \quad (2)$$

where q_t represents the amount (mg·g^{-1}) of calcium ions adsorbed at time t (min); C_0 (mg·L^{-1}) and C_t (mg·L^{-1}) are the initial concentration and t (min) concentration of calcium ions, respectively; V (L) is the volume of the calcium ions solution; and W is the weight of the RMCRs.

3. Results and Discussion

3.1. Characterization of the RMCRs

3.1.1. N$_2$ Adsorption–Desorption Isotherm Analysis

The pore structures and specific surface areas of the RMCRs were determined by using adsorption–desorption experiments, as shown in Figure 1a. The RMCRs' isothermal curves also showed type II curves with well-defined H$_1$ hysteresis-type loops, thus inferring cylindrical pores with a uniform macroporous structure [17]. The pore structures of the RMCRs had an average pore diameter of 42.40 nm (Figure 1b), Brunauer–Emmett–Teller (BET) surface area of 10.24 $m^2 \cdot g^{-1}$, and cumulative pore volume of 22.20 $mm^3 \cdot g^{-1}$. Thus, the RMCRs could exhibit excellent adsorption abilities on account of their high interconnect pores and high permeability, which facilitates the diffusion of adsorbents.

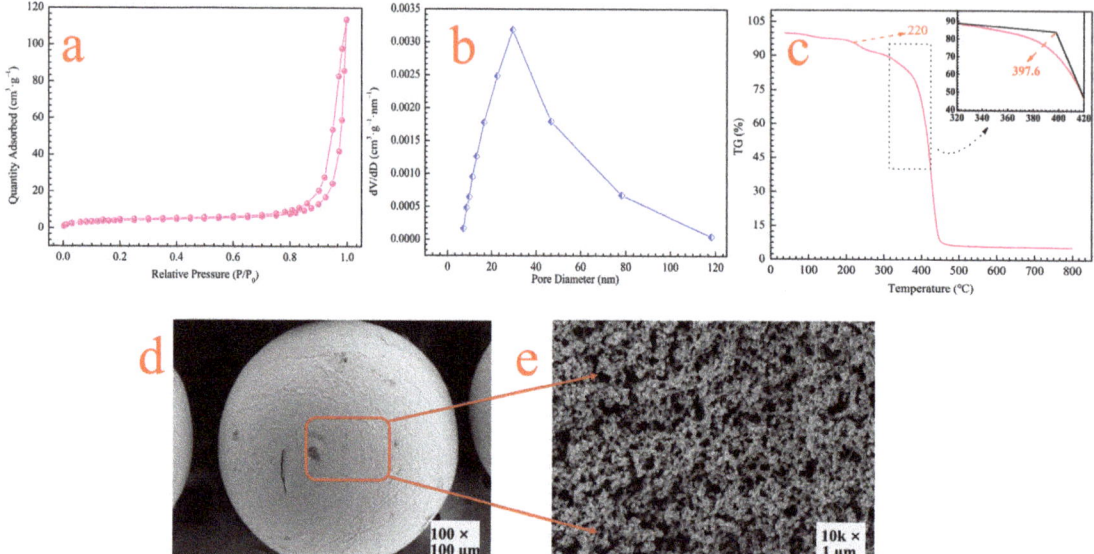

Figure 1. Characterization of the RMCRs by N_2 adsorption–desorption isotherms (**a**), pore size distributions (**b**), TGA (**c**), and SEM (**d**,**e**).

3.1.2. TGA Analysis

The thermal stability of the RMCRs was characterized by thermogravimetric analysis (STA449F3, Netzsch-Gerätebau GmbH, Selb, Germany). The results are presented in Figure 1c. RMCRs began to decompose at 220 °C due to the decomposition of rosin [18]. The onset temperature for RMCRs decomposition was approximately 220 [19]. The apparent weight loss of RMCR mainly occurred at temperatures ranging from 300 to 450 °C. The temperature of desalination in the sugar industry is commonly under 100 °C. Therefore, the RMCRs have high thermal and chemical stability and are appropriate for removing calcium ions from remelt syrup.

3.1.3. FE-SEM Analyses

The morphology and size of RMCRs were characterized by FE-SEM, and representative images are shown in Figure 1d,e. In the panoramic image, RMCRs were regular spheres with smooth, porous surfaces. The internal holes of the RMCR were interconnected. The rich porous structures of the RMCRs not only promoted the liquid mass transfer but also better access to the interaction sites [20]. Thus, it is beneficial to elevate the adsorption of calcium ions from remelt syrup [14].

3.2. Static Adsorption Experiments

3.2.1. Effect of RMCRs Dosage

The influences of RMCRs dosage on the adsorption capacity of calcium ions are presented in Figure 2a. With the increase in RMCRs dosage, the effective adsorption area increased, hence the removal efficiency promotion. However, the increase in the resin dosage increased the unsaturated loca on the adsorbent surface, thereby decreasing adsorption efficiency. In all the subsequent experiments, the solid-to-liquid ratio of 4.0 g·L^{-1} was selected as the optimal RMCRs dosage for calcium ions adsorption in consideration of efficiency and economy. Hence, RMCRs have great potential use in the removal of calcium ions from remelt syrup.

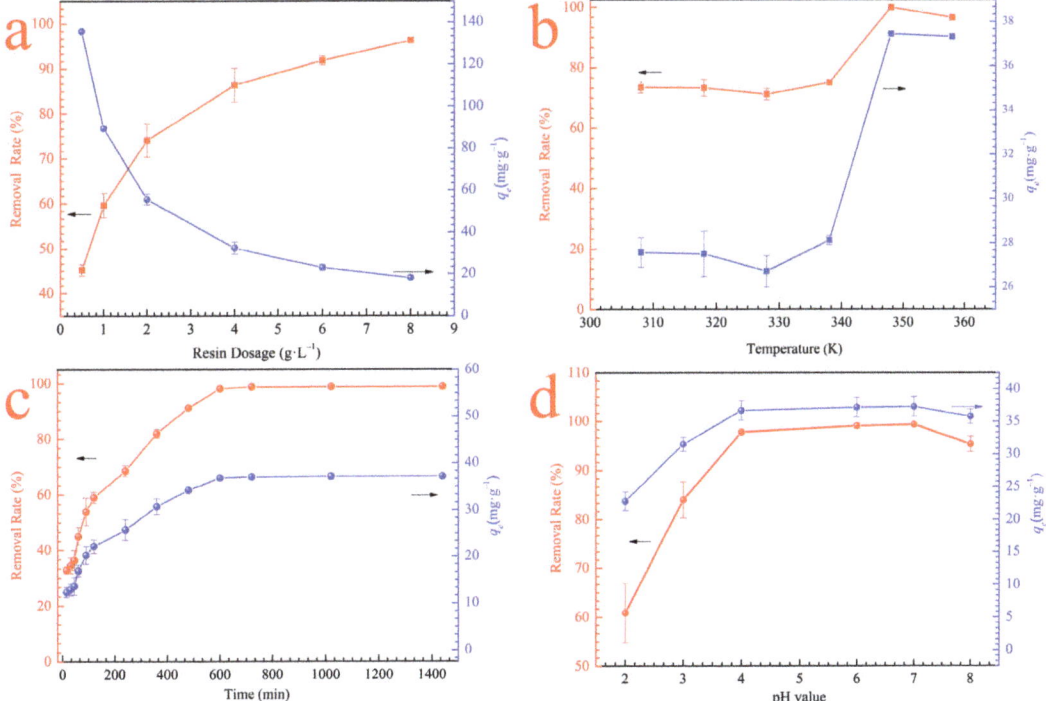

Figure 2. The adsorption effect of calcium ions adsorbed by RMCRs as a function of adsorbent dosage (**a**), temperature (**b**), contact time (**c**), and pH value of RMCRs (**d**).

3.2.2. Effect of Temperature

As scanned in Figure 2b, the temperature rose from 308 K to 348 K, and the q_e increased from 27.4 mg·g^{-1} to 37.5 mg·g^{-1} under other equal conditions. The increase in temperature enhanced the binding of calcium ions to RMCR adsorption sites, due to the movement of calcium ions in the solution being accelerated. Hence, 348 K was selected as the optimal temperature for calcium ions adsorption. The adsorption of calcium ions by RMCRs is an endothermic process.

3.2.3. Effect of Contact Time

According to Figure 2c, the calcium ions absorption capacity over RMCRs gradually increased in the beginning, and then steadily reached equilibrium after 600 min, which indicates a large number of active sites on the surface of the adsorbent in the initial stage. However, the adsorption sites were occupied, and the adsorption rate slowly decreased until equilibrium was reached. Adsorption equilibrium was reached in approximately 10 h, the q_e of RMCR was 37.05 mg·g^{-1}, and the corresponding removal rate was 90%. Thus, the contact time was set to 10 h in subsequent experiments.

3.2.4. Effect of pH

The pH plays a vital role in the adsorption of calcium ions by RMCRs. The effects of initial pH (2.0–8.0) on the calcium ions adsorption by the RMCRs are shown in Figure 2d.

The zeta potential value was correlated with the charge of the RMCRs and reflected their adsorption characteristics. Here, we focused on the main functional group of RMCRs involved in calcium ions adsorption, namely, the −COONa groups. This group could be very easily ionized to form −COO− groups. The zeta potentials of RMCRs are shown in Figure 3, with pH_{pzc} values of 2.7. At pH = 2 (pH < pH_{pzc}), the removal rate of calcium ions was lower, due to the positive charge on the surface of RMCRs; thus, the electrostatic repulsion interrupted the adsorption of calcium ions. When the solution pH = 3 (pH > pH_{pzc}), the surfaces of the RMCRs acquired negative charges. Thus, calcium ions and resins are attracted by electrostatic forces. Therefore, in the range of pH 2.0 to 3.0, the removal percentages sharply rose from 60.87% to 84.08%. When the pH > 4.0, the adsorption capacity was basically constant, as confirmed by the pH_{pzc} (zero charge points) and zeta potential of RMCRs. At pH 8.0, the adsorption capacity of RMCRs decreased because the system could present molecular agglomeration due to the number of intermolecular interactions and induced precipitation processes or an increase in viscosity [21]. The supreme adsorptivity for calcium ions was obtained at pH 7.0. Therefore, pH 7.0 is considered the optimum condition and used hereafter.

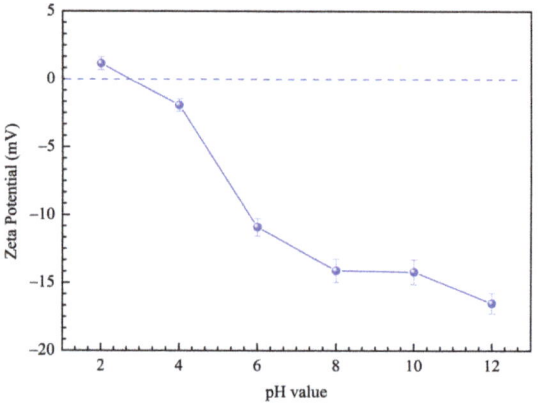

Figure 3. Zeta potentials of RMCRs at different pH.

3.3. Adsorption Kinetics

The absorption data were investigated by the pseudo-first-order kinetic, pseudo-second-order kinetic, and intraparticle diffusion equations to explore the calcium ions capacity of RMCRs. Linear forms of equations of these models can be written as follows:

Pseudo-first-order kinetic equation [22]:

$$\log(q_e - q_t) = \log q_e - \frac{K_1}{2.303} \cdot t \qquad (3)$$

Pseudo-second-order kinetic equation [23]:

$$\frac{t}{q_t} = \frac{1}{K_2 \cdot q_e^2} + \frac{t}{q_e} \qquad (4)$$

Intraparticle diffusion equation [8]:

$$q_t = K_3 \cdot t^{0.5} + C \qquad (5)$$

In the above equations, q_t and q_e are the calcium ions adsorption capacity (mg·g^{-1}) at time t and equilibrium time, respectively; K_1 (min^{-1}), K_2 (g·mg^{-1}·min^{-1}), and K_3 (mg·g^{-1}·min$^{0.5}$) are the rate constants of the pseudo-first-order kinetic, pseudo-second-order kinetic, and intraparticle diffusion equations, respectively; $t^{0.5}$ (min$^{0.5}$) is the square root of the contact time; and C represents the boundary layer thickness.

The fitting results are shown in Figure 4a–c, respectively. The pseudo-first-order equation explains the experimental data well compared with the other two equations, and the adsorbed quantities calculated (35.60 mg·g^{-1}) by this model are closer to those determined experimentally (37.05 mg·g^{-1}); thus, the main rate-limiting step of adsorption is physical adsorption. However, Figure 4c illustrates the separation of the adsorption process into two stages, namely fast adsorption and slow adsorption. Initially, calcium ions are adsorbed onto the surfaces of the RMCRs; after the surfaces are saturated, they gradually filter into the pore and inner surfaces of RMCRs by intraparticle diffusion until sorption decreases to equilibrium. Therefore, the calcium ions in aqueous solutions adsorbed onto RMCRs is a complicated procedure that involves boundary layer and intraparticle diffusion.

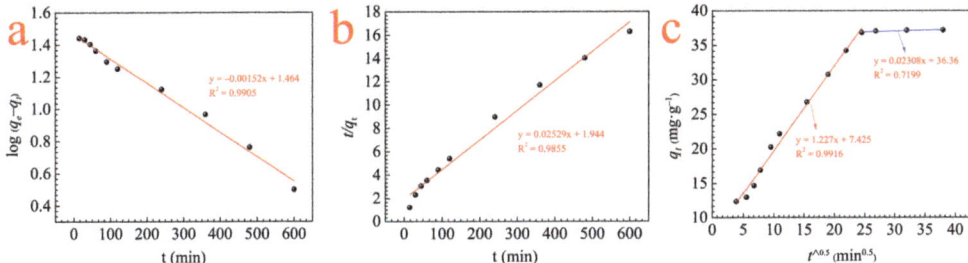

Figure 4. Behavior of RMCR adsorption of calcium ions: fits of the kinetic data of calcium ions adsorption on RMCRs using pseudo-first-order (**a**), pseudo-second-order (**b**), and intraparticle diffusion models (**c**).

3.4. Adsorption Isotherms and Thermodynamics

Adsorption isotherms facilitate the description of the interaction between calcium ions and RMCRs' surfaces at equilibrium. The adsorption isotherms at 328, 338, and 348 K are shown in Figure 5. The q_e of RMCRs increased with the concentration of calcium ions and initial temperature.

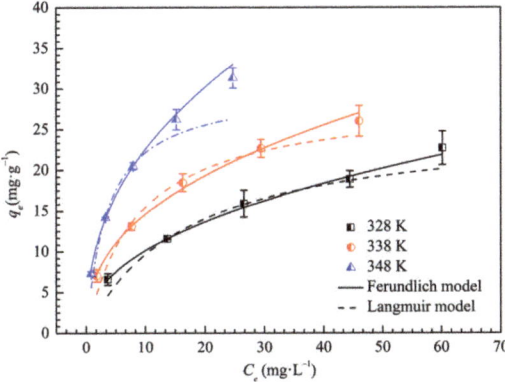

Figure 5. Isotherms for the adsorption of calcium ions on RMCRs at 328, 338, and 348 K.

The isotherm data were fitted by the Freundlich and Langmuir models to investigate the adsorption behavior. The experimental data of isotherm models are generally employed to delineate by adsorption equations and can be written as follows:

Freundlich model [22]:

$$q_e = k_F \cdot C_e^{\frac{1}{n}} \tag{6}$$

Langmuir model [24]:

$$q_e = \frac{q_m \cdot k_L \cdot C_e}{1 + k_L \cdot C_e} \quad (7)$$

where K_F (mg·L^{-1}) represents the Freundlich constant; $1/n$ is the strength of adsorption; and K_L (L·mg^{-1}) and q_m (mg·g^{-1}) are constants correlated with the affinity of the adsorption sites for RMCRs, respectively.

Table 1 summarizes the fitting results, in which the Freundlich and Langmuir models depict the isotherm data adequately. N is related to the adsorption driving force and the energy distribution of the adsorption sites. As shown in Table 1, the values of $1/n$ are between 0.1 and 0.5, which indicate that the adsorption of calcium ions on RMCRs is facile. As a result, the Freundlich model performed better ($R^2 > 0.99$) for describing the adsorption system in the range of concentrations and temperature ranges studied. Hence, calcium ions are adsorbed as a heterogeneous surface of an adsorbent and multilayer. Some heterogeneities on the surface of the RMCRs will take effect in calcium ions adsorption because of the existence of carboxyl and the abundant pore structures of RMCRs. These results further demonstrate the excellent promise in the removal of calcium ions by RMCRs in the sugar industry.

Table 1. Adsorption isotherm parameters of calcium ions onto RMCRs.

Temperature (K)	Freundlich Constants			Langmuir Constants		
	$1/n$	K_F (mg·g^{-1})	R^2	q_m (mg·g^{-1})	K_L (L·mg^{-1})	R^2
328	0.4353	3.586	0.9974	25.75	0.05989	0.9088
338	0.4056	4.735	0.9957	28.79	0.1158	0.9313
348	0.4280	5.716	0.9954	30.13	0.2768	0.9282

At 348 K, the adsorption capacity of calcium ions increases, illustrating the endothermic nature of the adsorption process.

Equations (8) and (9) were applied to the experimental results and K_d values of different temperatures to calculate the thermodynamic parameters (ΔH, ΔS, and ΔG) [25]:

$$\Delta G = -RT \ln K_d \quad (8)$$

$$\ln K_d = \frac{\Delta S}{R} - \frac{\Delta H}{RT} \quad (9)$$

where R is the universal gas constant (8.314 J·mol^{-1}·K^{-1}). lnK_d is plotted with $1/T$, the slope and intercept are acquired, and ΔH and ΔS are calculated; and the thermodynamic parameters are listed in Table 2. ΔG values (-1.729 to -3.133 kJ·mol^{-1}) were negative at various temperatures, proving that the process was feasible and spontaneous. In general, ΔG values between -20 and 0 kJ·mol^{-1} represent physisorption. The ΔH value (21.29 kJ·mol^{-1}) indicated that the process was endothermic. During calcium ions adsorption on RMCRs, the positive value of ΔS indicated that the randomness between the solid/solution interfaces increased.

Table 2. Thermodynamic parameters of the calcium ions adsorption onto RMCRs.

ΔH (kJ·mol^{-1})	ΔS (J·mol^{-1}·K^{-1})	ΔG (kJ·mol^{-1})		
		328 K	338 K	348 K
21.29	70.18	-1.729	-2.431	-3.133

3.5. Effect of Remelt Syrup Brix

The calcium ions removal at different Brix values of the remelt syrup was investigated. As shown in Figure 6a, the removal rate of calcium ions decreased at 55 °Bx. Brix values of the remelt syrup decreasing in remelt syrup can decrease the mass concentration gradient pressure and viscosity. Brix values of the remelt syrup provide a power to overcome the bulky transfer resistance of calcium ions between solution and RMCRs [26]. Hence, the reduction in the initial remelt syrup Brix can enhance the interaction strength between calcium ions and RMCRs. The reduction in the initial remelted syrup sugar content can gain the interaction of calcium ions with the sugar content of the remelted syrup. Considering adsorption efficiency, we selected 45 °Bx of remelt syrup in the subsequent experiment. On the other hand, it also shows that RMCRs can adapt to the viscosity and pressure of syrup and is suitable for the sugar industry.

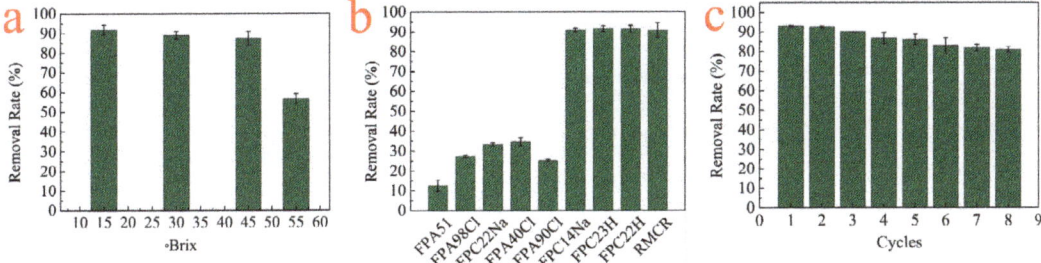

Figure 6. Effect of remelt syrup Brix on calcium ions adsorption onto RMCRs (**a**), comparison with various commercial resins (**b**), and reusability of RMCRs for calcium ions uptake (**c**).

3.6. Comparison with Various Commercial Adsorbents

The removal rate of calcium ions on the RMCRs was compared with those on other commercial resins (i.e., FPA 51, FPA 98 Cl, FPC 22 Na, FPA 40 Cl, FPA 90 Cl, FPC 14 Na, FPC 23 H, and FPC 22 H) (Figure 6b). The main physicochemical properties of the commercial resins are presented in Table 3. The removal rates of calcium ions by FPA 51, FPA 98 Cl, FPC 22 Na, FPA 40 Cl, FPA 90 Cl, FPC 14 Na, FPC 23 H, FPC 22 H, and RMCR were 12.54%, 27.26%, 33.25%, 34.61%, 25.15%, 90.82%, 91.65%, 91.48%, and 90.71%, respectively. FPC 14Na, FPC23 H, FPC22 H, and RMCRs revealed the superior adsorption abilities for calcium ions, suggesting that they would be excellent adsorbents for removing calcium ions from remelt syrup.

Table 3. Physicochemical properties of the commercial resins used.

Commercial Resins	Particle Size (mm)	Exchange Capacity (eq·L^{-1})	Matrix Structure	Functional Group
FPA51	0.49–0.69	≥1.3	Styrene-divinylbenzene copolymer	–NR$_2$
FPA98 Cl	0.63–0.85	≥0.8	polymethacrylic acid	R$_4$NOH
FPC22 Na	0.60–0.80	≥1.7	Styrene-divinylbenzene copolymer	–SO$_3$Na
FPA40 Cl	0.50–0.75	≥1.0	Styrene-divinylbenzene copolymer	R$_4$NOH
FPA90 Cl	0.65–0.82	≥1.0	Styrene-divinylbenzene copolymer	R$_4$NOH
FPC14 Na	0.60–0.80	≥2.0	Styrene-divinylbenzene copolymer	–SO$_3$Na
FPC23 H	0.58–0.80	≥2.2	Styrene-divinylbenzene copolymer	–SO$_3$H
FPC22 H	0.60–0.80	≥1.7	Styrene-divinylbenzene copolymer	–SO$_3$H
RMCR	0.35–0.83	≥0.3	polymethacrylic acid	–COONa

Lead adsorption capacities vary depending on adsorbent properties such as structure, surface area, porosity, and adsorbent polarity. Compared with FPA 51, FPA 98 Cl, FPA 40 Cl, and FPA 90 Cl are all anion exchange resins, and FPC 14 Na, FPC 23 H, FPC 22 H, and RMCRs are all cation exchange resins with superior adsorption abilities for calcium ions, thereby showing that the polarity of the adsorbent is a key element determining the adsorption capacity. Therefore, RMCRs with carboxyl functional groups are a potential adsorbent in the sugar industry.

3.7. Regeneration

One of the important factors in evaluating adsorbent performance is reusability. After the adsorption, RMCRs were regenerated with HCl (0.1 mol·L^{-1}) solutions, washed with deionized water until neutral, and used for the next adsorption experiments. The findings are shown in Figure 6c, and the regeneration efficiency was successively regenerated eight times. Even after eight regenerations, the RMCRs still contain a remarkable removal rate (80.87%). Hence RMCRs can be repeatedly used for the removal of calcium ions from remelt syrup.

3.8. Column Adsorption Performance and Models

Beyond the experiments already described, calcium ions from remelt syrup adsorption on RMCRs was estimated on fixed-bed columns. Figure 7 shows the relationship between C_t/C_0 and throughput volume when the bed depth is 4.0 cm. The breakthrough point is defined as the time when the effluent concentration reaches a percentage of the influent concentration (C_0), which is considered unacceptable, e.g., 10% ($C/C_0 = 0:1$). For $C/C_0 = 0.1$, the number of bed volumes that pass through the adsorbent was 480 BV. The Thomas [27], Yoon–Nelson [28], and Adams–Bohart models [29] are devoted to estimate the sorption adsorption behavior.

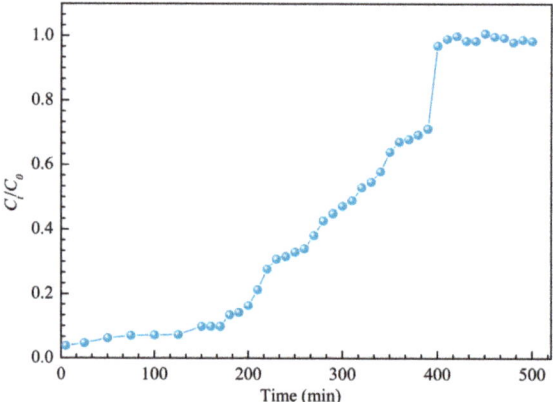

Figure 7. Breakthrough curve for calcium ions adsorption by RMCRs fixed columns.

The Thomas, Yoon–Nelson, and Adams–Bohart model equations can be written as follows, respectively:

Thomas model [27]:

$$\frac{C_t}{C_0} = \frac{1}{1 + \exp(\frac{K_{Th}q_0 m}{Q} - K_{Th}C_0 t)} \quad (10)$$

Yoon–Nelson model [28]:

$$\frac{C_t}{C_0} = \frac{\exp(K_{YN}t - \tau K_{YN})}{1 + \exp(K_{YN}t - \tau K_{YN})} \quad (11)$$

Adams–Bohart model [29]:

$$\frac{C_t}{C_0} = \exp(k_{AB}C_0 t - k_{AB}N_0 \frac{Z}{F}) \quad (12)$$

In the above equations, K_{Th}, K_{YN}, and k_{AB} are the Thomas rate constant (mL·min^{-1}·mg^{-1}), Yoon–Nelson rate constant (min^{-1}), and the Adams–Bohart rate constant (min^{-1}), respectively; q_0 represents the column adsorption ability (mg·g^{-1}), Q is the flow velocity (mL·min^{-1}), and m is the mass of the RMCR (g). C_0 and C_t are the calcium ion concentrations at the inlet and outlet, respectively (mg·L^{-1}). τ is the time (min) required for the adsorbate to breakthrough 50%. t (min) is the filtering time. N_0 is the saturation concentration of the bed (mg·L^{-1}), and t_b is the service time at breakthrough (h).

From Table 4, the breakthrough is more fitted with the Thomas model (R^2 = 0.966). According to Thomas model calculation, the dynamic saturated adsorption capacity of RMCRs for calcium ions from remelt syrup was 37.90 mg·g^{-1}. The calculated value was close to the actual experimental results (37.05 mg·g^{-1}). It was found that the Thomas model could be used to describe the dynamic adsorption characteristics of calcium ions adsorbed by RMCRs and predict the dynamic adsorption amount in industrial application.

Table 4. Model parameters for RMCR fixed-bed columns calcium ions.

Thomas Model			Yoon-Nelson Model			Adams-Bohart Model			
q_0 (mg·g^{-1})	$K_{Th} \times 10^{-5}$ (L·mg^{-1}·min^{-1})	R^2	$K_{YN} \times 10^{-2}$ (min^{-1})	τ (min)	R^2	Z (cm)	N_0 (mg·L^{-1})	$k_{AB} \times 10^{-5}$ (L·mg^{-1}·min^{-1})	R^2
37.90	6.440	0.9666	1.165	314.3	0.9666	4	484313	4.616	0.9638

3.9. Characterization of RMCRs before and after Adsorption of Calcium Ions

3.9.1. FTIR Analysis

The FTIR spectra of RMCRs and RMCRs with adsorbed calcium ions are shown in Figure 8. The spectra show wide absorption peaks at 3410 and 3391.52 cm^{-1}, which assign to the O−H bond stretching vibration in the hydroxyl function groups [30]. The bands at 2987 and 2937.93 cm^{-1} originate from the symmetry flex vibration of C−H bonds in −CH$_2$−, which are derived from the concatenation of carbonaceous species in RMCRs and remelt syrup. Wavenumbers indicate the adsorption of organic ingredients deposited on the RMCRs [31]. After calcium ions adsorption, the asymmetric −CH$_2$− stretching vibration shifts from 2987 cm^{-1} to 2937 cm^{-1}, thus indicating interactions with alkyl chains of RMCRs. The bands of calcium ions adsorbed on RMCRs at 1716.37 cm^{-1} (C=O), 1558.22 cm^{-1} (COO−), and 911.26 cm^{-1} (C−O−C) are assigned to the C=O in ester carboxyl or carboxyl groups, which are attributed to the carboxyl group on RMCRs. This proved the interaction of calcium ions and the –COO– of RMCRs. The wavenumbers at 1457.5 cm^{-1} (C−OH), 1418.3 cm^{-1} (COO−), and 1345.25 cm^{-1} (C−N) are due to the existence of proteins in the remelt syrup [32–35]. The sucrose compounds and phenols characteristic bands include 1052.26 (C−O) and 3391.52 cm^{-1} (O−H). These peaks are due to polysaccharides from remelt syrup [36]. The above results demonstrate that polysaccharides, protein, phenols, and sucrose can also be adsorbed on RMCRs [36].

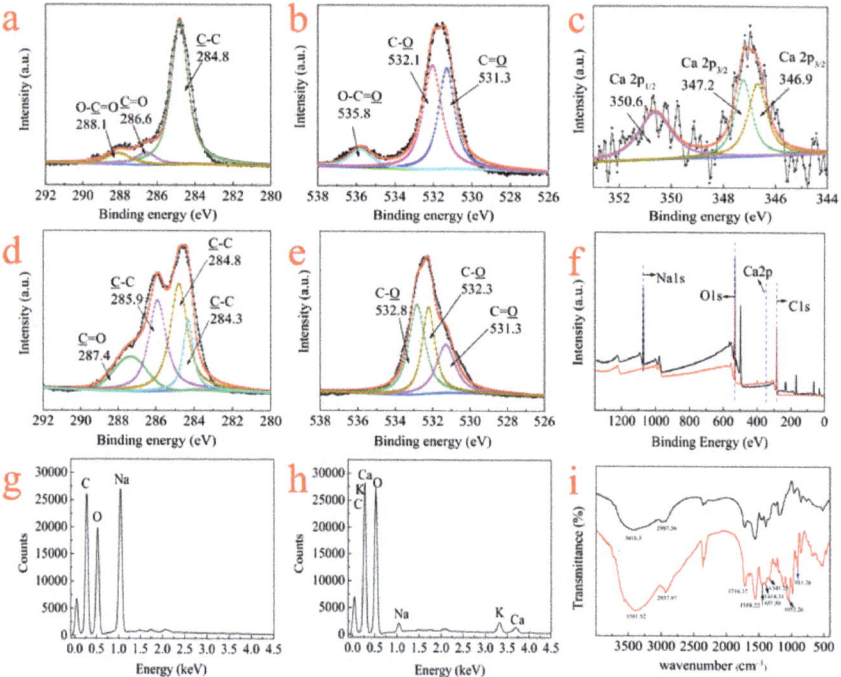

Figure 8. XPS C1s (**a**,**d**) and O1s (**b**,**e**) spectra of RMCRs and their deconvolution into component peaks before and after adsorption of calcium ions, respectively, and XPS spectra of calcium ions (**c**) for RMCRs with adsorbed calcium ions, XPS fully scanned spectra (**f**), EDS spectra of RMCRs (**g**) and RMCRs with adsorbed calcium ions (**h**), and FTIR spectra (**i**) of RMCRs (black solid line) and RMCRs with adsorbed calcium ions (red solid line).

3.9.2. XPS Analysis

XPS analysis was enforced to evaluate the elemental composites and chemical states RMCRs and RMCRs with adsorbed calcium ions, which are shown in Figure 8. The deconvolution of C 1s and O 1s peaks is also presented in Figure 8. The C 1s peaks of RMCRs (a) yield three contributions, which are 284.8 eV (C−C), 286.6 eV (C=O), and 288.1 eV (COO−) [37]. The shift in the carbon signal at 286.6 eV to lower BE after calcium ions adsorption is probably caused by the interaction of calcium ions with C−OH [38]. The peaks at 285.9 eV (C−C) and 287.4 eV (C=O) are related to sucrose compounds and phenols from remelt syrup. For the O 1s of RMCRs (b), the peak at 531.3 eV is attributable to C=O, that at 532.1 eV is attributable to C−O, and that at 535.8 eV is attributable to COO−. However, after calcium ions adsorption, the peak at 532.1 eV shifts to higher BE, which is caused by the interaction of calcium ions with oxygen atoms.

In Figure 8c, calcium ions are adsorbed through ionic bonding, thereby forming −(COO)$_2$Ca. The peaks of Ca 2p$_{3/2}$ at 347.2 and 346.9 eV represent the bonds between calcium ions and −COO−. The peak of Ca 2p$_{1/2}$ at 350.6 eV is attributable to CaCO$_3$ on the surfaces of the RMCRs. In Figure 8f, the Na 1s peak height of RMCRs with adsorbed calcium ions is lower than that of RMCRs; instead, the binding energy of Ca 2p at 346.6 eV is identified, which indicates that the ion exchange between −COONa and calcium ions contributes to calcium ions removal [39]. At the same time, it also shows that RMCRs have a certain ion exchange effect on the removal of calcium ions.

3.9.3. EDS Analysis

The energy-dispersive X-ray spectroscopy (EDS) analysis (Figure 8g) of RMCRs suggests that the RMCRs contain carbon, oxygen, and a mass of sodium. EDS analysis (Figure 8h) implies that RMCRs with adsorbed calcium ions contains a spot of sodium, calcium, and potassium. Carbon, oxygen, and sodium come from RMCRs; potassium and calcium are from remelt syrup. The EDS spectra of RMCRs and RMCRs with adsorbed calcium ions suggest that the ion exchange may drive the uptake process of calcium ions.

The comprehensive analysis of FTIR, XPS, and EDS showed that the calcium ions from remelt syrup were adsorbed on RMCRs in this work. RMCRs have a superior adsorption effect on calcium ions and have great potential for application in the sugar industry.

4. Conclusions

This work investigates the potential of RMCRs for calcium ions removal from remelt syrup. The results show that the maximum monolayer adsorption capacity of calcium ions is 37.05 mg·g^{-1} at a resin dosage of 4 g·L^{-1}, pH of 7.0, temperature of 75 °C, and contact time of 10 h. The removal rate of calcium ions from remelt syrup by RMCRs is 90.71%. The adsorption of calcium ions on RMCRs is pseudo-first-order in proportion and conforms to the Freundlich isotherm model. The adsorption process is endothermic, the adsorption process is physical adsorption and involves weak chemical bonds, and the analyses of FTIR, XPS, and EDS prove that ion exchange occurs during the adsorption process. The Thomas model describes the dynamic adsorption well. Compared with commercial resins, RMCRs have a superior removal rate for calcium ions from remelt syrup. In summary, RMCRs can be used as adsorbents for removal of calcium ions from remelt syrup, and potentially useful in improving the quality of remelt syrup and reducing or eliminating the use of chemicals in the sugar industry.

Author Contributions: Methodology, G.C.; formal analysis, G.C., W.L. and L.L.; investigation, G.C., W.L., L.L., F.L., X.L. and Q.H.; data creation, G.C.; writing—original draft preparation, G.C.; writing—review and editing, G.C., L.L. and X.L.; supervision, Q.H.; project administration, Q.H. and X.L. All authors have read and agreed to the published version of the manuscript.

Funding: This work was financially supported by the National Natural Science Foundation of China (31860192), the Natural Science Foundation of Guangxi (2020GXNSFBA159009), the China Postdoctoral Science Foundation (2020M683209), the Opening Project of Guangxi Key Laboratory of Clean Pulp & Papermaking and Pollution Control (2019KF28), and the Scientific Research Foundation of Guangxi Minzu University (2019KJQD10).

Institutional Review Board Statement: Not applicable.

Informed Consent Statement: Not applicable.

Data Availability Statement: The data presented in this study are available on request from the corresponding author.

Conflicts of Interest: The authors declare no conflict of interest.

References

1. East, C.P.; Fellows, C.M.; Doherty, W.O.S. Chapter 25-Scale in Sugar Juice Evaporators: Types, Cases, and Prevention. In *Mineral Scales and Deposits*; Elsevier (S&T): New York, NY, USA, 2015; pp. 619–637.
2. Li, W.; Ling, G.; Huang, P. Performance of ceramic microfiltration membranes for treating carbonated and filtered remelt syrup in sugar refinery. *J. Food Eng.* **2016**, *170*, 41–49. [CrossRef]
3. Phakam, B.; Moghaddam, L.; Baker, A.G. Compositional and structural changes of sugarcane evaporator deposits after concentrated sodium hydroxide treatment. *J. Food Eng.* **2017**, *214*, 1–9. [CrossRef]
4. Bakir, C.H.; Rackemann, D.W.; Doherty, W.O.S. Current perspective and future research directions on defecation clarification for the manufacture of raw sugar. *Sugar Ind.* **2021**, *146*, 634–642. [CrossRef]
5. Mouadili, H.; Majid, S.; Kamal, O. New grafted polymer membrane for extraction, separation and recovery processes of sucrose, glucose and fructose from the sugar industry discharges. *Sep. Purif. Technol.* **2018**, *200*, 230–241. [CrossRef]
6. Travaini, R.; Barrado, E.; Bolado-Rodríguez, S. Effect of ozonolysis pretreatment parameters on the sugar release, ozone consumption and ethanol production from sugarcane bagasse. *Bioresour. Technol.* **2016**, *214*, 150–158. [CrossRef] [PubMed]

7. Xiao, Y.; Lu, H.; Shi, C. High-performance quaternary ammonium-functionalized chitosan/graphene oxide composite aerogel for remelt syrup decolorization in sugar refining. *Chem. Eng. J.* **2022**, *428*, 132575. [CrossRef]
8. Li, W.; E, Y.; Cheng, L. Rosin-based polymer@silica core–shell adsorbent: Preparation, characterization, and application to melanoidin adsorption. *LWT* **2020**, *132*, 109937. [CrossRef]
9. Bednarczyk, P.; Irska, I.; Gziut, K. Novel Multifunctional Epoxy (Meth)acrylate Resins and Coatings Preparation via Cationic and Free-Radical Photopolymerization. *Ploymers* **2021**, *13*, 1718. [CrossRef]
10. Guo, L.; Lu, H.; Rackemann, D. Quaternary ammonium-functionalized magnetic chitosan microspheres as an effective green adsorbent to remove high-molecular-weight invert sugar alkaline degradation products (HISADPs). *Chem. Eng. J.* **2021**, *416*, 129084. [CrossRef]
11. Xie, W.; Li, H.; Sun, Y. Separating and purifying of Panax notoginseng saponins using a rosin-based polymer-bonded with silica as a high-performance liquid chromatography stationary phase. *Microchem. J.* **2022**, *176*, 107234. [CrossRef]
12. Ladero, M.; De Gracia, M.; Trujillo, F. Phenomenological kinetic modelling of the esterification of rosin and polyols. *Chem. Eng. J.* **2012**, *197*, 387–397. [CrossRef]
13. Liu, S.; Wang, J.; Huang, W. Adsorption of phenolic compounds from water by a novel ethylenediamine rosin-based resin: Interaction models and adsorption mechanisms. *Chemosphere* **2019**, *214*, 821–829. [CrossRef] [PubMed]
14. Li, P.; Qin, L.; Wang, T. Preparation and adsorption characteristics of rosin-based polymer microspheres for berberine hydrochloride and separation of total alkaloids from coptidis rhizoma. *Chem. Eng. J.* **2020**, *392*, 123707. [CrossRef]
15. Li, P.; Wang, T.; Lei, F. Preparation and evaluation of paclitaxel-imprinted polymers with a rosin-based crosslinker as the stationary phase in high-performance liquid chromatography. *J. Chromatogr. A* **2017**, *1502*, 30–37. [CrossRef]
16. Li, H.; Song, X.; Li, P. Separation of alkaloids and their analogs in HPLC using rosin-based polymer microspheres as stationary phases. *New J. Chem.* **2021**, *45*, 6856. [CrossRef]
17. Jiang, Y.; Wang, Z.; Zhou, L. Highly efficient and selective modification of lignin towards optically designable and multifunctional lignocellulose nanopaper for green light-management applications. *Int. J. Biol. Macromol.* **2022**, *206*, 264–276. [CrossRef]
18. Nirmala, R.; Woo-Il, B.; Navamathavan, R. Influence of antimicrobial additives on the formation of rosin nanofibers via electrospinning. *Colloids Surf. B: Biointerfaces* **2013**, *104*, 262–267. [CrossRef]
19. Jiang, Y.; Wang, Z.; Liu, X. Highly Transparent UV-Shielding, and Water-Resistant Lignocellulose Nanopaper from Agro-Industrial Waste for Green Optoelectronics. *ACS Sustain. Chem. Eng.* **2020**, *8*, 17508–17519. [CrossRef]
20. Zhao, H.; Gao, W.C.; Li, Q. Recent advances in superhydrophobic polyurethane: Preparations and applications. *Adv. Colloid Interface Sci.* **2022**, *303*, 102644. [CrossRef]
21. Palacio, D.A.; Vásquez, V.; Rivas, B.L. Chromate ion removal by water-soluble functionalized chitosan. *Polym. Adv. Technol.* **2020**, *32*, 2690–2699. [CrossRef]
22. Liang, J.; He, Q.; Zhao, Y. Synthesis of sulfhydryl modified bacterial cellulose gel membrane and its application in adsorption of patulin from apple juice. *LWT* **2022**, *158*, 113159. [CrossRef]
23. Ho, Y.S.; McKay, G. Pseudo-second order model for sorption processes. *Process Biochem.* **1999**, *34*, 451–465. [CrossRef]
24. Langmuir, I. The adsorption of gases on plane surfaces of glass, mica and platinum. *J. Am. Chem. Soc.* **1918**, *40*, 1361–1368. [CrossRef]
25. De Mattos, N.R.; De Oliveira, C.R.; Camargo, L.G.B. Azo dye adsorption on anthracite: A view of thermodynamics, kinetics and cosmotropic effects. *Sep. Purif. Technol.* **2019**, *209*, 806–814. [CrossRef]
26. Zhu, H.Y.; Jian, R.; Xiao, L. Preparation, characterization, adsorption kinetics and thermodynamics of novel magnetic chitosan enwrapping nanosized gamma-Fe_2O_3 and multi-walled carbon nanotubes with enhanced adsorption properties for methyl orange. *Bioresour. Technol.* **2010**, *101*, 5063–5069. [CrossRef]
27. Thomas, H.C. Heterogeneous ion exchange in a flowing system. *J. Am. Chem. Soc.* **1944**, *66*, 1664–1666. [CrossRef]
28. Yoon, Y.H.N.; Nelson, J.H. Application of gas adsorption kinetics–II. A theoretical model for respirator cartridge service life and its practical applications. *Am. Ind. Hyg. Assoc. J.* **1984**, *45*, 517–524. [CrossRef]
29. Bohart, G.S.; Adams, E.Q. Some aspects of the behavior of charcoal with respect to chlorine. *J. Am. Chem. Soc.* **1920**, *42*, 523–544. [CrossRef]
30. Lei, Y.; Wan, Y.; Zhong, W. Phosphonium-Based Porous Ionic Polymer with Hydroxyl Groups: A Bifunctional and Robust Catalyst for Cycloaddition of CO_2 into Cyclic Carbonates. *Polymers* **2020**, *12*, 596. [CrossRef]
31. Luo, W.; Lu, H.; Lei, F. Structural elucidation of high-molecular-weight alkaline degradation products of hexoses. *Food Sci. Nutr.* **2020**, *8*, 2848–2853. [CrossRef]
32. Sun, Y.; Li, P.; Wang, T. Alkaloid purification using rosin-based polymer-bonded silica stationary phase in HPLC. *J. Sep. Sci.* **2019**, *42*, 3646–3652. [CrossRef] [PubMed]
33. Wang, T.; Li, P.; Sun, Y. Camptothecin-imprinted polymer microspheres with rosin-based cross-linker for separation of camptothecin from Camptotheca acuminata fruit. *Sep. Purif. Technol.* **2020**, *234*, 116085. [CrossRef]
34. Li, W.; Hang, F.; Li, K. Development and Application of Combined Models of Membrane Fouling for the Ultrafiltration of Limed Sugarcane Juice. *Sugar Tech* **2018**, *21*, 524–526. [CrossRef]
35. Cheng, B.-X.; Gao, W.-C.; Ren, X.-M. A review of microphase separation of polyurethane: Characterization and applications. *Polym. Test.* **2022**, *107*, 107489. [CrossRef]

36. Li, W.; Ling, G.; Lei, F. Ceramic membrane fouling and cleaning during ultrafiltration of limed sugarcane juice. *Sep. Purif. Technol.* **2018**, *190*, 9–24. [CrossRef]
37. Gao, W.C.; Wu, W.; Chen, C.Z. Design of a Superhydrophobic Strain Sensor with a Multilayer Structure for Human Motion Monitoring. *ACS Appl. Mater. Interfaces* **2022**, *14*, 1874–1884. [CrossRef] [PubMed]
38. Wang, A.; Zhu, Q.; Xing, Z. Design and synthesis of a calcium modified quaternized chitosan hollow sphere for efficient adsorption of SDBS. *J. Hazard. Mater.* **2019**, *369*, 342–352. [CrossRef] [PubMed]
39. Carrera, K.; Huerta, V.; Orozco, V. Formation of vacancy point-defects in hydroxyapatite nanobelts by selective incorporation of Fe^{3+} ions in Ca(II) sites. A CL and XPS study. *Mater. Sci. Eng. B* **2021**, *271*, 115308. [CrossRef]

MDPI
St. Alban-Anlage 66
4052 Basel
Switzerland
www.mdpi.com

Polymers Editorial Office
E-mail: polymers@mdpi.com
www.mdpi.com/journal/polymers

Disclaimer/Publisher's Note: The statements, opinions and data contained in all publications are solely those of the individual author(s) and contributor(s) and not of MDPI and/or the editor(s). MDPI and/or the editor(s) disclaim responsibility for any injury to people or property resulting from any ideas, methods, instructions or products referred to in the content.

www.ingramcontent.com/pod-product-compliance
Lightning Source LLC
LaVergne TN
LVHW070608100526
838202LV00012B/596